UNITEXT for Physics

UNITEXT for Physics series publishes textbooks in physics and astronomy, characterized by a didactic style and comprehensiveness. The books are addressed to upper-undergraduate and graduate students, but also to scientists and researchers as important resources for their education, knowledge, and teaching.

Guido Boffetta • Angelo Vulpiani

Probability in Physics

Foundations and Applications

Guido Boffetta
Department of Physics
University of Torino
Torino, Torino, Italy

Angelo Vulpiani
Dipartimento di Fisica
Università degli Studi di Roma La Sapienza
Roma, Italy

ISSN 2198-7882 ISSN 2198-7890 (electronic)
UNITEXT for Physics
ISBN 978-3-032-10406-9 ISBN 978-3-032-10407-6 (eBook)
https://doi.org/10.1007/978-3-032-10407-6

This Springer imprint is published by the registered company Springer Nature Switzerland AG
The registered company address is: Gewerbestrasse 11, 6330 Cham, Switzerland

Preface

The first reaction to this book might be:

Why another book on probability? Why this title?

A perhaps not original answer is that probability is both a language and a technical and conceptual tool—well established today and of undeniable importance. Just to keep it short, we can quote the great James Clerk Maxwell:

The true logic for this world is the calculus of Probabilities.

Although *the calculus of Probabilities* (the theory of probability, in modern language) is present in almost all fields of physics, from our experience we know that very often physics students have serious gaps (both technical and conceptual) even on basic aspects of probability. Paradoxically, even those who deal with statistical mechanics are sometimes not immune to these shortcomings in preparation. In our opinion, these deficiencies are due to the educational organization, which typically relegates the presentation—often fragmentary and using only elementary mathematics—of the basic concepts and techniques of probability to the early years in laboratory courses and a few mentions in the statistical mechanics course. Meanwhile, a systematic and more rigorous treatment is available only in advanced (non-mandatory) courses, usually with a mathematical or mathematical-physics approach.

Even on fundamental topics such as the law of large numbers and the central limit theorem, one often encounters vague—or even incorrect—ideas about the true scope of these results. Among many, we can mention the ridiculous statement (attributed to Poincaré[1]) that circulates about the widespread presence of the Gaussian function in many phenomena, shrouded in an unnecessary aura of mystery: *Experimentalists think that it is a mathematical theorem, while the mathematicians believe it to be an experimental fact.*

[1] We refuse to believe that the great scientist could have said, except perhaps as a joke, such a nonsense.

This widespread disinterest in probability among physicists is in some respects inexplicable and almost paradoxical, since the modern developments of probability theory were clearly inspired by physics. Even without being experts in the history of science, one can easily argue that in the second half of the nineteenth century, the old classical approach to probability had no chance of development, both due to internal problems, but above all due to the lack of serious applications. It was precisely the stimuli coming from physics, starting with the development of statistical mechanics by J.C. Maxwell and L. Boltzmann, and Brownian motion (with A. Einstein, M. Smoluchowski, and P. Langevin) that enabled the modern development of the theory of probability and of stochastic processes.

In physics, the theory of probability has a central role, and this for several reasons. In addition to the obvious ones (data analysis), we can list:

- **(a)** Clarifying some fundamental aspects of statistical mechanics. For example: the mathematical meaning of statistical ensembles, the importance of the many degrees of freedom involved in macroscopic objects, the principle of maximum entropy (so often cited inappropriately)
- **(b)** Understanding the apparent dichotomy between the deterministic description (in terms of differential equations) of classical physics and the use of probabilistic approaches
- **(c)** Navigate in the modeling of "complex" phenomena, for example, those involving degrees of freedom with very different characteristic times

The structure of the book is as follows:

The **First Part** (Chaps. 1, 2, and 3) is a *General Introduction to Probability*. Particular emphasis is given to conditional probability, marginal densities, and limit theorems (law of large numbers, central limit theorem, and theory of large deviations). Some examples are introduced with the explicit purpose of highlighting how many results of statistical mechanics are nothing but applications of general aspects of the theory of probability.

In the **Second Part** (Chaps. 4, 5, and 6), we present the *Fundamental Concepts of Stochastic Processes*. After a discussion of Brownian motion, we introduce Markov chains and stochastic processes whose probability density is governed by the Fokker-Planck equations. Two brief digressions, on the Monte Carlo method and the use of stochastic differential equations for climate models, give an idea of the practical importance of stochastic processes in physics.

The **Third Part** (Chaps. 7, 8, and 9) is a *Selection of Advanced Topics*: analysis of deterministic chaotic systems in probabilistic terms; generalization of the central limit theorem for variables with infinite variance (Lévy stable functions); relevance of non-Gaussian distributions in diffusion processes. Finally, we discuss some of the many aspects of entropy, from statistical mechanics, to information theory, to deterministic chaos. Needless to say, the choice of topics in this third part is largely dictated by the interests of the authors.

For completeness, in each chapter we have included some exercises and proposed simple numerical experiments.

The **Appendix** is divided into two parts. The first contains a syllabus of definitions and basic concepts, placed at the end of the book so as not to burden the main text with material that may be superfluous for some readers. In the second part, we discuss three interesting topics, even if somewhat marginal to physics: an application (technically elementary but with non-trivial consequences) of probability to genetics; the statistics of extreme events; and an application (at the edge of what is permissible) of probability to the distribution of prime numbers.

Finally, the **solutions** to the exercises are provided.

The prerequisites required of the reader are only basic university-level mathematics, namely, derivatives, integrals, series, elementary combinatorial calculus, and Fourier transforms.

The first acknowledgment goes to Luca Peliti, who encouraged us in this project with suggestions and advice that improved the book. Stefano Berti, Massimo Cencini, Fabio Cecconi, Filippo De Lillo, Massimo Falcioni, Giacomo Gradenigo, Miguel Onorato, Davide Vergni, and Dario Villamaina read parts of the text and suggested improvements; to all of them, our thanks. A special thanks to Alessandro Sarracino, who found many unclear points, typos, and errors.

Turin, Italy | Guido Boffetta
Rome, Italy | Angelo Vulpiani
Summer 2025

Competing Interests The authors have no competing interests to declare that are relevant to the content of this manuscript.

Contents

Chapter 1
Introduction

1.1 A Bit of History: The Dawn

As usual, we begin with a brief historical excursus, obviously without any claim to completeness or methodological rigor. In our case, the historical introduction is not only a tribute to the founding fathers but also an opportunity to highlight some conceptual aspects.

Probability theory is among the most recent mathematical disciplines, formalized only in the twentieth century. It emerged much later (compared to branches such as geometry, algebra, and analysis), and initially for rather frivolous reasons: games of chance. The origin of the theory of probability is often attributed to a problem raised by Chevalier A.G. de Méré (an avid gambler) and solved by B. Pascal. The problem was as follows: to explain why (as empirical evidence shows) betting on getting at least one 6 in 4 rolls of a (fair) die is more likely to win than lose, while betting on getting at least one double 6 in 24 rolls of a pair of dice is more likely to lose than win. De Méré was convinced that the probabilities of winning in the two cases should be equal.[1] Pascal's answer was to correctly calculate the probabilities and thus obtain the right result. The solution is rather simple: since the die has 6 equivalent faces, the probability of getting a 6 in one roll is $p = 1/6$, the probability of getting a result different from 6 is $q = 1 - p = 5/6$, so the probability of not getting any 6 in 4 rolls is $q^4 = (5/6)^4$ and the probability of getting at least one 6 in 4 rolls is $1 - q^4 = 1 - (5/6)^4 = 671/1296 \simeq 0.517$.

[1] For the curious reader, de Méré's (incorrect) argument was as follows: if in a repeated game the probability of winning in a single attempt is $1/N$, then there will be a number n^* (proportional to N) such that if $n > n^*$, the probability of winning by betting on at least one event in n trials is greater than the probability of losing. In the first game $N = 6$ while in the second $N = 36$, so by taking $n = 4$ and $n = 24$ respectively, the ratio n/N is $4/6 = 24/36$, so it should not be possible that in the first game the probability of winning is greater than that of losing, while in the second it is the opposite.

G. Boffetta, A. Vulpiani, *Probability in Physics*, UNITEXT for Physics,
https://doi.org/10.1007/978-3-032-10407-6_1

For the second game, the procedure is similar: the probability of a double 6 in a roll of a pair of dice is $p = 1/36$, so the probability of getting a different result is $q = 1 - 1/36 = 35/36$, and the probability of getting at least one pair of 6 in 24 rolls is $1 - q^{24} = 1 - (35/36)^{24} \simeq 0.491$. It is remarkable that de Méré, though a modest mathematician, was able to detect from observation a difference of only 3%.

The first "serious" work (which anticipates one of the fundamental results of the theory of probability, namely the law of large numbers) is the treatise *Ars conjectandi* by Jakob Bernoulli, published posthumously in 1713.

1.1.1 Probability as Frequency

The law of large numbers assures us that, in a sense that we will discuss in detail in Chap. 3, if P is the probability that a certain event occurs (for example, getting heads in a coin toss), then if we perform N (independent) trials, denoting by N^* the number of times the event occurs, then "apart from rare events" we have that

$$\lim_{N\to\infty} \frac{N^*(N)}{N} = P. \tag{1.1}$$

More precisely, for every $\epsilon > 0$ the probability that $N^*(N)/N$ deviates by more than ϵ from P becomes arbitrarily small as N increases:

$$\lim_{N\to\infty} P\left(\left|\frac{N^*(N)}{N} - P\right| > \epsilon\right) = 0. \tag{1.2}$$

A proposal that sounds natural is to define the probability of the event as its frequency in the limit of many trials. Strictly speaking, this is not without flaws and one could object that:

(a) there is a circularity (self-reference) in (1.2);
(b) there is the problem that (1.2) does not guarantee the convergence of N^*/N to P for all sequences, but only for "almost all sequences"; for example, in the coin toss, the sequence with N heads can obviously occur;
(c) how large must N be?
(d) how do we decide that the approximation is good?

It is not easy to answer these questions in elementary terms. Answers, at least partial ones, will be discussed when we deal with limit theorems and the theory of large deviations.

1.1.2 Classical Probability

In 1716 A. de Moivre in *Doctrine de Chances* introduced the so-called classical definition of probability: the probability of an event is the ratio between the number of favorable cases and the possible ones, assuming that all events are equiprobable.[2] Furthermore, de Moivre shows, in a specific situation, a particular case of the central limit theorem.[3]

A fundamental role for classical probability is played, at the beginning of the nineteenth century, by P.S. Laplace with his *Théorie analytique des probabilités*. In the 1814 edition, the essay *Essai Philosophique des probabilites* is included, which contains the famous manifesto on determinism:

We must therefore consider the present state of the universe as the effect of its previous state and as the cause of its future state. An intelligence that, for a given instant, knew all the forces by which nature is animated and the respective situation of the beings that compose it, if in addition it were deep enough to master these data, would embrace in the same formula the movements of the largest bodies of the universe and of the lightest atom: nothing would be uncertain for it and the future, like the past, would be present to its eyes.

It may seem paradoxical to find what is considered the quintessential formulation of the deterministic point of view, in a work dedicated to probability.

A few pages later Laplace tries to give an explanation for the apparent contradiction between the existence of irregular phenomena and determinism:

The curve described by a single molecule of air or vapor is regulated with the same certainty as the planetary orbits: there is no difference between them, except that which is placed there by our ignorance. Probability is relative in part to this ignorance, in part to our knowledge.

The classical definition of probability, which is based on discrete events, has evident difficulties in the case where continuous variables are considered. However, the approach can be generalized, at least in certain situations, and lead to geometric probability. To give an example, let us consider the following problem: a room is paved with square tiles of side L, a coin of diameter $d < L$ is tossed, and we ask for the probability (which, note well, has not yet been defined) that the coin falls straddling at least 2 tiles. In Fig. 1.1 the area is shown in which the center of the coin must fall to have the desired event. It is natural (or at least it seems so) to suppose that the probability is the ratio between the area of the shaded part and the area of the tile, thus $p = 1 - (L - d)^2/L^2$. Therefore, in the context of geometric

---

[2] A minimum of reflection leads one to suspect that there is a weak point in this definition because the concept of equiprobable is self-referential.

[3] As noted by Kac and Ulam, for some purists de Moivre's result would not be considered very profound since it is only a rather simple application of elementary combinatorial formulas and Stirling's approximation $n! \simeq \sqrt{2\pi n}\, n^n e^{-n}$.

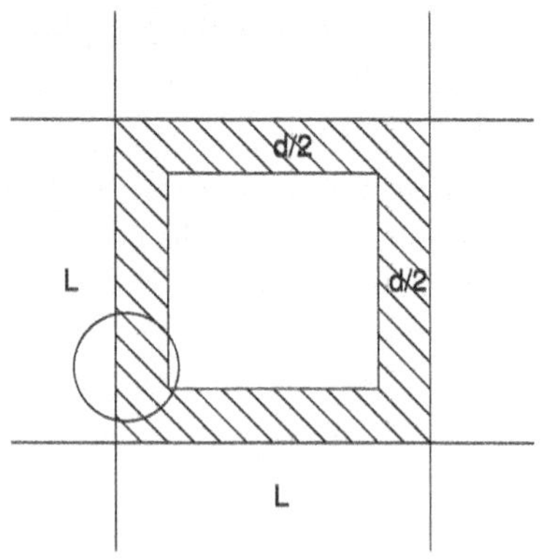

Fig. 1.1 The problem of tossing a coin. The shaded area, of width $d/2$, represents the region in which the center of the coin must fall so that it straddles at least two tiles

probability, probability is defined as the ratio between the area of the favorable event and the total area.[4] Obviously, in one dimension, instead of area, length is used, and in three dimensions, volume.

At first glance everything seems reasonable, unfortunately, as we will see later, the underlying idea of geometric probability hides a subtle aspect that cannot be overcome without a radical rethinking of the problem on solid mathematical foundations.

How to Calculate π by Tossing Needles on the Floor

An interesting application of geometric probability is the so-called Buffon's needle problem . A floor is covered with parquet with identical strips of width d, a needle of length $L < d$ is dropped at random. The question is the probability that the needle falls straddling two strips. Let us denote by x the distance from the line separating two strips and the center of the needle, and by θ the angle that the needle makes with the line perpendicular to the separator, see Fig. 1.2, and assume that x and θ are uniformly distributed in $(0, d]$ and $(0, \pi/2]$ respectively. There is an intersection, see Fig. 1.2, when

$$\frac{L}{2}\cos\theta > x, \tag{1.3}$$

this corresponds to the needle intersecting the left separator line, or

$$\frac{L}{2}\cos\theta > d - x, \tag{1.4}$$

[4] Note that, in modern terms, in the problem of tossing the coin, the definition is equivalent to assuming that the joint probability density of the variables x and y (respectively the abscissa and ordinate of the center of the coin) is constant, and thus one is still in fact using the concept of "equiprobable" from classical probability.

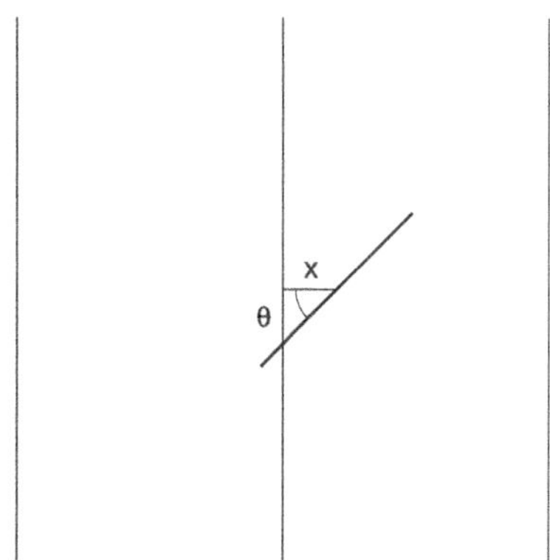

Fig. 1.2 Buffon's needle problem with length L thrown at random onto a floor made up of strips of width d

corresponding to the intersection with the line on the right. Therefore, the probability is given by

$$P = \frac{Area(F)}{Area(E)},$$

where $Area(F)$ and $Area(E)$ are respectively the area of the region for which (1.3) or (1.4) holds, and the total area. A straightforward calculation gives

$$P = \frac{4}{\pi d} \int_0^{\frac{\pi}{2}} \frac{L}{2} \cos\theta \, d\theta = \frac{2L}{\pi d}. \tag{1.5}$$

At this point, invoking the law of large numbers, one can think of calculating π by throwing a needle a large number of times ($N \gg 1$) and counting the number of times (N^*) that the needle crosses the dividing line. This yields a good estimate of $P \simeq N^*/N$ and thus, from (1.5), an estimate of π:

$$\pi \simeq \frac{2LN}{dN^*} .$$

This result can be considered the ancestor of the Monte Carlo method: by resorting to the law of large numbers and using a stochastic methodology, it is possible to solve a problem (the calculation of π) that has nothing random about it.

1.1.3 Bertrand's Paradox

Colloquial language is not always able to avoid paradoxical situations, as is clearly evident in the following example (due to Bertrand). Consider the problem: given a circle of unit radius, draw a random chord. Calculate the probability that the length of the chord is greater than $\sqrt{3}$ (the side of the inscribed equilateral triangle).

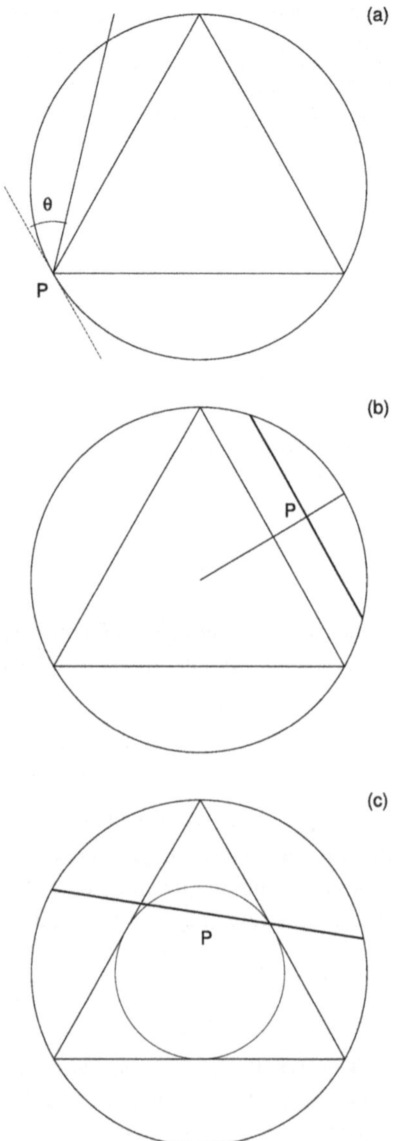

Fig. 1.3 Three possible solutions to Bertrand's problem

First Answer take a point P on the edge of the disk. All chords starting from P are parameterized by an angle θ, see Fig. 1.3a. If the chord is to be longer than $\sqrt{3}$, the angle θ must be within a sector of $60°$ out of an interval of 180, so the probability is $60/180 = 1/3$.

Second Answer consider a point P on a radius and the chord passing through P and perpendicular to the radius, see Fig. 1.3b. The chord is longer than $\sqrt{3}$ if its center P is in the inner part (of length $1/2$), so since the radius is 1, the probability is $1/2$.

Third Answer if the center of the chord falls within the disk of radius $1/2$, then the chord is longer than $\sqrt{3}$, see Fig. 1.3c, since the area of this circle is $\pi/4$ while the total area is π, the probability is $1/4$.

Which is the correct answer? Simply, the question is ill-posed, because "draw a random chord" is decidedly too vague, and in each of the three answers there is a hidden assumption that seems natural, but is actually arbitrary. In the first, it is assumed that the angle θ is uniformly distributed; in the second, that the center of the chord is uniformly distributed along the diameter; while in the third, that the center of the chord is uniformly distributed within the circle.

It is clear that Bertrand's paradox shows the ambiguity of some apparently intuitive ideas, which are often invoked (inappropriately) in the physical sciences. For example, it makes no sense, without some specific argument dictated by physics or otherwise, to say *it is natural to assume that a probability density is uniform.*

1.2 The Theory of Probability Becomes a Mature Science

By the end of the nineteenth century, it was already clear that the theory of probability required a profound and solid systematization, both technical and conceptual. It is no coincidence that D. Hilbert, in his famous address at the International Congress of Mathematicians in 1900 in Paris, in the list of 23 open problems, included (as the sixth) the problem of *treating in an axiomatic way those parts of the physical sciences in which mathematics plays an important role; in particular, the theory of probability ..... and the rigorous and satisfactory development of the method of averages in mathematical physics, in particular in the kinetic theory of gases.*

The initiator of this project was E. Borel, who realized that Lebesgue's measure theory should be the mathematical foundation of probability theory. The formalization program it can be considered concluded in 1933 with the publication of A.N. Kolmogorov's book *Grundbegriffe der Wahrscheinlichkeitsrechnung* (Foundations of the Theory of Probability).[5] Let us briefly discuss the axioms introduced by Kolmogorov to formalize the theory of probability and their meaning.

Let us consider a set Ω of elementary events ω and let $\mathcal{F}$ be a family of subsets of Ω. We call Ω the sample space and the random events (or simply events) the elements of $\mathcal{F}$:

[5] Kolmogorov's work can be seen as the final summa that summarizes the long process of systematization and formalization that involved many mathematicians, among whom (in addition to Borel and Kolmogorov) F.P. Cantelli, M. Fréchet, A. A. Khinchin, P. Lévy, and M. von Mises.

I $\mathcal{F}$ is an algebra of sets, that is, $\Omega \in \mathcal{F}$, and $\mathcal{F}$ is closed under the operations of union, intersection, and complement, that is, if $A \in \mathcal{F}$ and $B \in \mathcal{F}$, then $A \cap B$, $A \cup B$, and $\overline{A} = \Omega - A$ are also contained in $\mathcal{F}$.[6]

II To each element A of $\mathcal{F}$ is associated a non-negative real number (the probability of A) $P(A)$.

III $P(\Omega) = 1$.

IV If two sets A and B are disjoint (i.e., $A \cap B = \emptyset$), then $P(A \cup B) = P(A) + P(B)$.

The triple $(\Omega, \mathcal{F}, P)$ is called a probability space. It is an easy exercise to show that

$$P(\overline{A}) = 1 - P(A), \ \ P(\emptyset) = 0, \ \ 0 \leq P(A) \leq 1.$$

Let us now discuss the conceptual (and empirical) meaning of Kolmogorov's four axioms, which is important if the theory of probability is to be not only a branch of mathematics but also usable in the sciences.

Let us consider an "experiment" $\mathcal{S}$ that can be repeated a practically unlimited number of times, and let us consider a given group of possible events as the outcome of the realization of experiment $\mathcal{S}$.

Axiom **I** specifies the "objects" for which it makes sense to define probability. For example, if $\mathcal{S}$ consists of tossing a pair of distinguishable coins, then the elementary events are the visible faces of the two coins, so $\Omega = \{HH, HT, TH, TT\}$ where HT indicates heads for the first coin and tails for the second, and so on.[7]

The properties of the probability of an event $P(A)$ must be such that:

(a) it is practically certain that if $\mathcal{S}$ is repeated a very large number of times ($N \gg 1$) and the event A occurs M times, then M/N is very close to $P(A)$;

(b) if $P(A)$ is very small, then it is practically certain that event A does not occur in a single realization of $\mathcal{S}$.

Since $0 \leq M/N \leq 1$ and for the event Ω we always have $M = N$, axioms **II** and **III** are natural.

If A and B are incompatible (i.e., A and B are disjoint), then $M = M_1 + M_2$ where M, M_1, and M_2 are respectively the number of times the events $A \cup B$, A, and B occur, so $M/N = M_1/N + M_2/N$, which suggests axiom **IV**.

In the particularly important case where the elementary event ω is a real number, then Ω is the real number line R, and the natural choice for $\mathcal{F}$ is the set of half-open intervals $[a, b)$. It is convenient to introduce the distribution function:

$$F(x) = P([-\infty, x)) \ ,$$

[6] $B - A$ is the set containing the elements of B but not those of A, so $\overline{A} = \Omega - A$ consists of the elements not contained in A.

[7] "Obviously," if the coins are not biased, we have $P(HH) = P(HT) = P(TH) = P(TT) = 1/4$.

that is, the probability that the random variable X is less than x, and the probability density

$$p_x(x) = \frac{dF(x)}{dx} ,$$

obviously we have

$$P([a, b)) = \int_a^b p_x(x')dx' .$$

The notation $p_x(x)$ or $p_X(x)$ indicates that the density refers to the random variable X. To be rigorous, the definition of probability density makes sense only if $F(x)$ is differentiable; however, if we accept that $p_x(x)$ can be a generalized function (for example, with Dirac deltas), the problem does not arise.[8] For the toss of a fair die we have $F(x) = 0$ for $x < 1$, $F(x) = 1/6$ for $1 \leq x < 2$, $F(x) = 2/6$ for $2 \leq x < 3$, etc., and therefore

$$p_x(x) = \sum_{n=1}^{6} \frac{1}{6}\delta(x - n). \tag{1.6}$$

We also note that Kolmogorov's axioms are perfectly compatible with the definition of classical probability and geometric probability; moreover, the set of axioms is not contradictory.[9] Let us add that Kolmogorov was a convinced frequentist in the sense that he believed that the interpretation of probability in terms of frequency provided the best connection between the mathematical formalism and physical reality.

1.2.1 The Concept of Independence

Two events A and B are said to be independent if

$$P(A \cap B) = P(A)P(B), \tag{1.7}$$

more generally, $A_1, A_2, \ldots, A_N$ are independent if

$$P(A_1 \cap A_2 \cap \ldots \cap A_N) = \prod_{k=1}^{N} P(A_k). \tag{1.8}$$

---

[8] If the random variable is discrete, then $F(x)$ is piecewise constant.

[9] It is enough to consider the case in which the only possible event is Ω, so $\mathcal{F}$ consists only of Ω and $\emptyset$ and furthermore $P(\Omega) = 1$, $P(\emptyset) = 0$.

This definition sounds rather intuitive; however, given the importance of the concept, it is appropriate to reinforce the intuition. The probability of $A \cap B$ if A and B are independent must be a function only of $P(A)$ and $P(B)$:

$$P(A \cap B) = F(P(A), P(B)), \tag{1.9}$$

we now need to determine the form of $F(x, y)$. Let us consider the following experiment: tossing a coin, suitably biased so that the probability of getting heads is p, and a die with four faces numbered from 1 to 4, also biased so that the faces 1, 2, 3, and 4 appear with probabilities p_1, p_2, p_3, and p_4 respectively (obviously $p_1 + p_2 + p_3 + p_4 = 1$). We assume that the toss of the coin and the roll of the die are two independent events and consider the event $T \cap (1 \cup 2)$, that is, getting heads (event A) and getting either the side numbered 1 or the one numbered 2 (event B). From axiom **IV** we have $P(B) = p_1 + p_2$ and therefore from (1.9) we have

$$P(T \cap (1 \cup 2)) = F(p, p_1 + p_2), \tag{1.10}$$

where $p = P(A)$. Since $T \cap (1 \cup 2) = (T \cap 1) \cup (T \cap 2)$ and the events $T \cap 1$ and $T \cap 2$ are disjoint, by axiom **IV** and (1.8) we have

$$P(T \cap (1 \cup 2)) = F(p, p_1) + F(p, p_2)\,.$$

Therefore, $F(x, y)$ must satisfy the equation

$$F(x, y_1 + y_2) = F(x, y_1) + F(x, y_2). \tag{1.11}$$

At this point, noting that $F(1, y) = y$ and $F(x, 1) = x$, assuming (which seems natural) that $F(x, y)$ is continuous in x and y, from (1.11) we obtain $F(x, y) = xy$.

Another argument to "convince oneself" of (1.7): suppose that in $N \gg 1$ trials the event A occurs $N(A)$ times, B occurs $N(B)$ times, and $A \cap B$ occurs $N(A \cap B)$ times. We can write

$$\frac{N(A \cap B)}{N} = \frac{N(A \cap B)}{N(B)} \frac{N(B)}{N},$$

at this point, if A and B are independent, it makes sense to assume that the occurrence of B does not influence the occurrence of A, and therefore for large N, $N(A \cap B)/N(B)$ should not be different from $N(A)/N$; now, identifying frequencies with probabilities, (1.7) follows.

1.2.2 Another Axiom

Kolmogorov adds a fifth axiom (apparently innocent), that of continuity or **countable additivity**

V If $\{A_j\}$, with $j = 1, 2, \ldots$ is a countable collection of pairwise disjoint events in $\mathcal{F}$, then

$$P(\bigcup_{j=1}^{\infty} A_j) = \sum_{j=1}^{\infty} P(A_j).$$

To be precise, in his 1933 book Kolmogorov introduced an equivalent axiom: if $\{A_j\}$ is a decreasing sequence of events such that $A_1 \supseteq A_2 \supseteq ..$ with $\lim_{N\to\infty} \bigcap_{j=1}^{N} A_j = \emptyset$ then $\lim_{N\to\infty} P(A_N) = 0$.

At this point, the attentive reader will have realized the mathematical structure that lies behind Kolmogorov's axioms: we are in the presence of measure theory in a suitable "disguise." Let us recall, for completeness (and convenience), that a non-negative function of A, $\mu(A)$, is called a measure if the following properties hold:

Property 1 If $A_1, A_2, \ldots$ are disjoint and measurable sets, then their union $A_1 \cup A_2 \cup \ldots$ is also measurable and

$$\mu(A_1 \cup A_2 \cup \ldots) = \mu(A_1) + \mu(A_2) + \ldots$$

Property 2 If A and B are measurable and $A \subset B$, then the set $B - A$ is measurable and, by Property 1, we have $\mu(B - A) = \mu(B) - \mu(A)$.

Property 3 A certain set E has measure 1: $\mu(E) = 1$.

Property 4 If two measurable sets are congruent (for example, superimposable by rotations and/or translations), they have the same measure.

M. Kac summarized Kolmogorov's approach with the slogan *probability theory is measure theory plus a soul*. The soul is the notion of statistical dependence, and the mathematical tool that quantifies this notion is conditional probability.

Countable additivity is a delicate assumption. As Kolmogorov explicitly admits, it is hardly possible to explain its empirical meaning, since in the description of any experimentally observable random process we can only obtain finite probability spaces. With axiom **V** (which in measure theory corresponds to the property of σ-additivity, or countable additivity) we are in fact deciding to (arbitrarily) limit the theory to a subclass of models.[10]

[10] The importance of assuming, or not, axiom **V** is well known in the context of measure theory. For example, in 1905 G. Vitali provided an example of a subset of the real line that is not measurable with respect to any measure that is positive, translation-invariant, and σ-additive (in particular, the

1.3 Probability and the Real World

Beyond the technical aspects, it is natural to ask why a physicist should be concerned with probability. After all, throwing dice or coins on tiled floors is not exactly an exciting activity, even if it is not trivial.

There are at least three interesting situations in physics in which the use of the theory of probability seems inevitable:[11]

(a) The number of degrees of freedom involved in the problem is large, and one is not interested in the details of the system but only in the collective behavior of a few variables. This is the case of statistical mechanics (see Chaps. 7 and 9).
(b) There is not complete control over the initial conditions of the system, which, even if deterministic, exhibits "irregular" (unstable) behavior; this is the case of systems with deterministic chaos, which can manifest regardless of the number of degrees of freedom (see Chap. 7).
(c) In a "complex" phenomenon[12] one does not have control over the "many causes at play"; the paradigmatic example is Brownian motion, in which a colloidal particle (much larger than the molecules of the fluid in which it is immersed) moves in an irregular way. In the presence of a clear separation of scale between the characteristic time of the molecules and that of the colloidal particle, it is possible to describe the phenomenon with a stochastic process (Langevin equation, see Chaps. 4 and 6).

The cases listed above have a common element: for some reason, there is not complete knowledge of the phenomenon, and one "settles" for a non-detailed description, that is, not of the complete state of the system but of a projection or a "coarse-grained" description. This way of conceiving probability in physics is very old, and can also be found in Laplace, who maintained that probability is relative to our ignorance.

Using the terminology of the philosophy of science, one can say that this is an epistemic conception of probability, which is seen as something merely linked to "our ignorance" and not to the intrinsic nature of classical systems (which are deterministic).[13]

Lebesgue measure). For the construction of the Vitali set, the axiom of choice (Zermelo's axiom) is indispensable, a delicate and controversial aspect of set theory. For example, for Lebesgue, the existence of the non-measurable set found by Vitali is not acceptable for "empiricists" who reject the axiom of choice.

[11] Throughout this book we will not deal with quantum phenomena, which are described by a formalism in which probability plays a fundamental role, even more so than in the classical context.

[12] The terminology is intentionally vague.

[13] On the contrary, at least in the orthodox interpretation of the Copenhagen school, in quantum mechanics probability has an ontic character, that is, intrinsic to the phenomenon and not dependent on the lack of information of the observer.

This is not the place for a detailed epistemological analysis. For those readers who are concerned about the apparently "negative" meaning of epistemic (since it is not linked to the true nature of the system, but only to the limited human capabilities), we point out that epistemic is by no means synonymous with subjective; moreover, the ontic and epistemic levels often intersect in subtle ways. As an example, we can mention the case of chaotic deterministic dynamical systems, whose deterministic nature is certainly an ontic property, as it is intrinsic to the system and independent of the skill of the scientist studying the system. On the other hand, the limited predictability, and more generally the chaotic nature, is to be considered an epistemic property; in fact, in the calculation of Lyapunov exponents and Kolmogorov-Sinai entropy, one resorts to a description of the initial state that is not arbitrarily accurate (for example, for Kolmogorov-Sinai entropy, a partition of phase space into cells of finite size is introduced). However, this does not mean that chaos is not an objective phenomenon.[14]

Often, the terms "probability," "statistical result," and similar expressions are understood as something vague and imprecise, opposed to certainty. The hope for absolute certainties (in sciences other than mathematics), if it ever truly existed (apart from some simplistic popularizations), has long since faded. J.C. Maxwell wrote in 1873:

The fact that the same antecedents are followed by the same consequences is a metaphysical doctrine. No one can deny it. But it is not very useful in the world in which we live, where the same antecedents never occur and nothing ever happens exactly the same way twice. In fact, as far as we know, one of the antecedents could be the precise date and location of the event; in this case, our experience would be completely useless. [...] The axiom of physics that, in a certain sense, has the same nature is "that similar antecedents are followed by similar consequences." But here we have moved from "equality" to "similarity," from absolute accuracy to a more or less rough approximation.

The use of a probabilistic approach therefore seems inevitable: the austere Maxwell became the champion of probability theory, a science that, born for *trivial and immoral* reasons (in Maxwell's own words), has become the necessary guide for interpreting the world.

One might observe that we have spoken of probability but without an unassailable and explicit definition. The classical and geometric definitions have their own problems, and even the viewpoint in terms of frequencies of very long series does not seem immune to criticism. In Kolmogorov's axiomatic approach, probability is simply not defined; only the properties that probabilities must have are stated.

[14] Denoting by ϵ the size of the partition cells, the Kolmogorov-Sinai entropy, which measures the degree of unpredictability of the system, becomes independent of ϵ in the limit of small ϵ (see Chap. 9).

As for a possible definition, let us consider the following questions in common language that refer to the concept of probability:

(a) what is the probability that Roma will win the championship in the next soccer season?
(b) what is the probability that the government will fall next week?
(c) what is the probability of getting all heads in 50 coin tosses?
(d) what is the probability that a helium molecule in a gas at a temperature of 400 K and a pressure of one atmosphere has a speed greater than 100 m/s?

    There is no unanimous consensus that all four questions are well-posed. In (a) and (b), we have single (non-repeatable) events, and therefore the interpretation of probability in terms of frequencies must be excluded, and one must try to define probability as the "degree of belief" that an individual holds. In the subjectivist approach, developed especially by B. de Finetti, a definition of probability is proposed that is applicable to random experiments whose elementary events are not considered equally possible and that are not necessarily repeatable an arbitrary number of times under the same conditions. The probability of an event is defined as the maximal price that a rational individual considers fair to pay to receive 1 if the event occurs, 0 if the event does not occur. To make the definition concretely applicable, a criterion of coherence is added: probabilities must be assigned in such a way that it is not possible to obtain a certain win or loss. Within the subjectivist approach, all statements (a), (b), (c), and (d) make sense.

On the contrary, Kolmogorov is very clear in stating that not for all events can a probability be defined: *The assumption that a definite probability exists for a given event under certain conditions is a hypothesis that must be verified and justified in each individual case.*

Once it has been decided that it makes sense to speak of the probability of event X, the problem remains of answering the question *what is the probability that event X will occur?* One must seek an answer case by case. In the paradigmatic example of the fair coin, the hypothesis that the probability of heads is $1/2$ follows from obvious symmetry considerations, which, of course, are not always possible.[15]

We adopt the point of view that only questions of the type (c) and (d) are scientifically relevant, assuming (at least as a working hypothesis) the validity of the frequentist interpretation. Despite all its limitations, this way of understanding probability has a great physical advantage: frequencies contain objective, experimentally verifiable information that has nothing to do with our beliefs or knowledge about nature.

We report two more quotations, with which we fully agree:

One of the most important tasks of probability theory is to identify those events whose probability is close to zero or one (A.A. Markov); *The entire epistemo-*

[15] How can one determine the probability of heads if the coin is biased? Is it possible to do so without resorting to the "empirical" method of many tosses?

logical value of probability theory is based on this: random phenomena, when considered in their collective action on a large scale, generate a non-random regularity (B.V. Gnedenko and A.N. Kolmogorov).

A very important aspect for physics is the close relationship between the frequentist point of view and the foundations of statistical mechanics (in particular, the ergodic problem). For L. Boltzmann (and also A. Einstein), the probability of an event is nothing other than the percentage of time during which the event occurs; this idea allows for an unambiguous definition of probability even in a single system, without invoking statistical ensembles.

We conclude the chapter with a very brief discussion on the claim that probabilistic statements would not be falsifiable. On this point, Popper is very clear: *Probabilistic estimates are* ***not*** *falsifiable. And, of course, they are not even verifiable ...experimental results, no matter how numerous and favorable, can never definitively establish that the relative frequency of "heads" is* $1/2$ *and will always be* $1/2$.

Taken literally, these statements are obviously true, although it is natural to ask:

(a) do we really conduct research to falsify theories?
(b) can anyone, in good faith, claim that the Maxwell-Boltzmann distribution for the velocities of gas molecules (a probabilistic statement) has not been "verified" (directly or indirectly) beyond any reasonable doubt?
(c) it is true that we will never know the "exact" value of the probability of a certain event, but this is a fact that is not particularly disturbing. How can one be sure that the length of an object is exactly 22.33 cm?
(d) are the problems raised by Popper regarding probability not common to all sciences? Are we not facing the old (unsolvable) problem of the justification of induction?

We are inexorably sliding toward the much broader question of the connection between the theories we construct, the interpretation of empirical observations, and physical reality. In agreement with Reichenbach, it seems acceptable to us that, if we believe that statements about the physical world have meaning, we can just as confidently believe in the meaning of the concept of probability. For now, we find no better escape than a quote from A. Einstein:

When the laws of Nature refer to reality, they are not certain. When they are certain, they do not refer to reality.

Exercises

1.1 Prove that:

$$P(A \cup B) = P(A) + P(B) - P(A \cap B) \tag{1.12}$$

$$P(A \cup B \cup C) = \tag{1.13}$$

$$= P(A) + P(B) + P(C) - P(A \cap B) - P(A \cap C) - P(C \cap B) + P(A \cap B \cap C).$$

1.2 Given two events A and B such that $P(A) = 3/4$ and $P(B) = 1/3$, show that $P(A \cap B) > 1/12$.

1.3 Show that there cannot exist two sets A and B such that $P(A) = 4/10$, $P(B) = 3/10$, and $P((\Omega - A) \cap (\Omega - B)) = 2/10$.

1.4 Find the probabilities of winning in the lottery for a single number, an ambo (pair), a terno (triple), etc. Recall that in the lottery game, 5 numbers are drawn from a container holding 90 numbers.

Recommended Readings

For the history of the modern theory of probability:

J. von Plato, *Creating Modern Probability* (Cambridge University Press, Cambridge, 1994)

For a brief discussion on probability and measure theory:

M. Kac, S. Ulam, *Mathematics and Logic* (Dover Publications, New York, 1992)

A must-read to deepen conceptual aspects:

A.N. Kolmogorov, *Grundbegriffe der Wahrscheinlichkeitsrechnung* (1933); English translation *Foundations of the Theory of Probability* (Chelsea Publ. Comp. 1956). Available at: http://www.kolmogorov.com/Foundations.html

For an introduction to the various interpretations of probability:

A. Hájek, Interpretations of Probability, in *Stanford Encyclopedia of Philosophy*. Available at: http://www.seop.leeds.ac.uk/entries/probability-interpret

Chapter 2
Some Results with a Bit of Formalism

2.1 Conditional Probability

One of the most important concepts in all of probability theory is certainly that of conditional probability. Knowing how to use it correctly allows one to avoid falling into insidious traps (often presented as paradoxes). We can say that the first level of understanding of probability theory is reached if one is able to correctly use conditional probability.

If $P(B) > 0$ then the probability of A given B is:

$$P(A|B) = \frac{P(A \cap B)}{P(B)}. \tag{2.1}$$

The motivation for the previous formula can be understood by resorting to the classical interpretation of probability: let N be the number of possible outcomes and let N_A, N_B, and N_{AB} denote the number of times event A, B, and $A \cap B$ occur, respectively. Then

$$P(A|B) = \frac{N_{AB}}{N_B} = \frac{N_{AB}}{N}\frac{N}{N_B}$$

since $P(B) = N_B/N$ and $P(A \cap B) = N_{AB}/N$, we obtain (2.1).

As an example, consider the roll of a fair die: let B be the set of odd numbers $B = \{1, 3, 5\}$ and A the number 1. The probability of getting 1 knowing that the result is odd is

$$P(A|B) = P(1|B) = \frac{1/6}{1/2} = \frac{1}{3},$$

G. Boffetta, A. Vulpiani, *Probability in Physics*, UNITEXT for Physics,
https://doi.org/10.1007/978-3-032-10407-6_2

whereas if A is the number 2 we have

$$P(A|B) = P(2|B) = 0\ ,$$

in agreement with intuition.

Note that if A and B are independent then $P(A|B) = P(A)$ (the converse also holds); in this case, knowing that event B has occurred does not change the information about A and therefore from (2.1) $P(A \cap B) = P(A)P(B)$.

By writing in (2.1) $P(A \cap B) = P(B \cap A) = P(B|A)P(A)$, it is immediate to obtain:

$$P(B|A) = P(A|B)\frac{P(B)}{P(A)}, \tag{2.2}$$

a relation known as the **Bayes' formula**.

Let us derive a couple of useful results in which conditional probability plays a relevant role.

Theorem of Total Probability

Let $\{B_i\}_{i=1}^N$ be a partition of Ω, that is, $B_i \cap B_j = \emptyset$ if $i \neq j$ and $\cup_{i=1}^N B_i = \Omega$, then

$$P(A) = \sum_{i=1}^{N} P(A|B_i)P(B_i). \tag{2.3}$$

Multiplication Theorem

$$P(A_1 \cap A_2 \cap \ldots \cap A_N) = P(A_1)P(A_2|A_1)P(A_3|A_2 \cap A_1) \times$$
$$P(A_4|A_3 \cap A_2 \cap A_1) \ldots .. P(A_N|A_{N-1} \cap A_{N-2} \cap \ldots \cap A_2 \cap A_1) \tag{2.4}$$

To prove (2.3), just observe that the events $C_i = A \cap B_i$ are independent, so $P(\cup_i C_i) = \sum_i P(C_i)$, moreover, since $\cup_i B_i = \Omega$ we have $\cup_i C_i = A$, thus $P(A) = \sum_i P(C_i)$. At this point, from the identity $P(C_i) = P(A \cap B_i) = P(B_i)P(A|B_i)$, (2.3) follows.

Equation (2.4) is obtained by repeatedly using the definition of conditional probability:

$$P(A_1 \cap A_2 \cap \ldots \cap A_N) = P(A_N|A_{N-1} \cap A_{N-2} \cap \ldots \cap A_2 \cap A_1) \times$$

$$P(A_{N-1} \cap A_{N-2} \cap \ldots \cap A_2 \cap A_1) = P(A_N | A_{N-1} \cap A_{N-2} \cap \ldots \cap A_2 \cap A_1) \times$$
$$P(A_{N-1} | A_{N-2} \cap \ldots \cap A_2 \cap A_1) P(A_{N-2} \cap \ldots \cap A_2 \cap A_1)$$

and so on.

The birthday problem is a fun exercise that can be easily solved with the help of (2.4). Given N people ($N < 365$), determine the probability that at least 2 were born on the same day (we ignore leap years and assume that births are uniformly distributed throughout the year). Let A denote the event *at least* 2 *people were born on the same day*, and $\overline{A}$ the complementary event *there are no pairs of people born on the same day*, then

$$P(A) = 1 - P(\overline{A}) \, ,$$

let A_1 denote the event *the second person was not born on the same day as the first*, A_2 the event *the third person was not born on the same day as the first and the second*, and so on. From the multiplication theorem we have

$$P(\overline{A}) = P(A_1) P(A_2 | A_1) \ldots P(A_{N-1} | A_1 \cap \ldots \cap A_{N-2})$$

A moment of reflection convinces us that

$$P(A_1) = \left(1 - \frac{1}{365}\right) \, , \quad P(A_2 | A_1) = \left(1 - \frac{2}{365}\right) \, ,$$

$$P(A_3 | A_1 \cap A_2) = \left(1 - \frac{3}{365}\right) \quad \ldots$$

$$P(A_{N-1} | A_1 \cap \ldots \cap A_{N-2}) = \left(1 - \frac{N-1}{365}\right).$$

Therefore,

$$P(A) = 1 - \prod_{j=1}^{N-1} \left(1 - \frac{j}{365}\right) \, . \tag{2.5}$$

An approximate formula can be obtained by noting that

$$\prod_{j=1}^{N-1} \left(1 - \frac{j}{365}\right) = \exp \sum_{j=1}^{N-1} \ln \left(1 - \frac{j}{365}\right) \simeq$$

$$\simeq \exp\left(-\sum_{j=1}^{N-1} \frac{j}{365}\right) = \exp\left(-\frac{N(N-1)}{730}\right). \tag{2.6}$$

From (2.5) for $N = 5$ we have $P = 0.027$, for $N = 10$, $P = 0.117$; $N = 20$, $P = 0.411$; $N = 22$, $P = 0.476$; $N = 23$, $P = 0.507$; $N = 60$, $P = 0.994$; $N = 64$, $P = 0.997$. So already with 23 people there is a probability greater than 1/2 that at least two people share the same birthday, and with 60 there is "almost certainty," a result that is not intuitive.

2.1.1 Fake Paradoxes: Just Know How to Use Conditional Probability

Many of what are often presented as paradoxes in probability theory arise from a lack of understanding of the concept of conditional probability. The most trivial example (but often with dramatic consequences) is that of overdue numbers in the lottery. In newspapers (even those considered serious) and on television networks (including public ones, which should not propagate manifestly incorrect ideas) great emphasis is often given to the fact that a certain number (say 21) has not come up on a given wheel for a large number of draws (say 150). The (incorrect) conclusion is that in the next draw the appearance of 21 should be "almost certain" because (this is the incorrect argument) "it is unlikely that a number will not come up for 151 consecutive times." Obviously, this is not the case, and many unwary players have ruined themselves by following these follies.[1]

The solution is elementary: one must not confuse:

(A) P = the probability of not having the number 21 come up for 151 consecutive times
(B) $\widetilde{P}$ = the probability that 21 does not come up after it has not come up for 150 times.

Since the probability of a single number being drawn is $1/18$, and the draws are independent, in the first case we have

$$P = \left(1 - \frac{1}{18}\right)^{151} = \left(\frac{17}{18}\right)^{151} \simeq 0.000178.$$

[1] Regarding the social consequences of gambling, recall that already Laplace, at the beginning of the nineteenth century, unsuccessfully tried to convince the French parliament of the immorality of the lottery (state-run). One could suggest that the house (ultimately the State) should behave as with the sale of cigarettes (on the packages of which phrases such as *Smoking causes lung cancer* are printed, which is true but rather vague). In the case of the lottery, it would be even easier and more objective: it would be enough to write that the probabilities of winning by betting on a single number, ambo, terno, quaterna, and cinquina are respectively 1/18; 1/400.5; 1/11748; 1/511038 and 1/43949268 while the house pays respectively 11.23; 250; 4500; 120,000 and 6,000,000, so the player would have all the objective data and could decide whether to play or not. For further information on the social consequences of gambling, see the interesting website www.fateilnostrogioco.it.

In the second case, knowing that 21 has not come up for 150 times is irrelevant (since the draws are independent), so $\widetilde{P} = 1 - 1/18 = 17/18 \simeq 0.9444$, a very different result!

Bayes' Formula in Medicine

In the 1990s, J. Tooby and L. Cosmides (two influential psychology researchers) discussed an interesting experiment in which a group of doctors and Harvard Medical School students were asked the following question

A disease has an incidence rate of $1/1000$. *There is a test that can detect its presence. This test has a false positive rate of* 5%. *An individual takes the test. The result is positive. What is the probability that the individual is actually sick?*

The correct answer, which can be easily obtained from Bayes' formula, is about 2%. Only 18% of the participants gave the correct answer, while as many as 58% answered that the probability was 95%. The fact that the majority gave the (incorrect) answer $P(M|p) = 95\%$ is "understandable": the (wrong) reasoning followed would be the following.[2] *In a population of* 100,000 *individuals there are about* 100 *positives, but there is a* 5% *error, so the number of true positives is about* 95, *and therefore the probability sought is about* 95%.

Here is the actual solution. Let $P(M) = 0.001$ be the probability of being sick, $P(S) = 1 - P(M) = 0.999$ the probability of being healthy, $P(p|S) = P_e = 0.05$ is the probability of a false positive, that is, testing positive while being healthy, and $P(n|M)$ the probability of a false negative, that is, testing negative while being sick. For simplicity, we assume[3] $P(n|M) = P(p|S) = P_e$. The probability we are looking for is $P(M|p)$, using Bayes' formula:

$$P(M|p) = P(p|M)\frac{P(M)}{P(p)},$$

since $P(p|M) = 1 - P(n|M) = 1 - P_e$ and, by the law of total probability, $P(p) = P(p|S)P(S) + P(p|M)P(M) = P_e(1 - P(M)) + (1 - P_e)P(M)$ we obtain

$$P(M|p) = \frac{(1 - P_e)P(M)}{P_e(1 - P(M)) + (1 - P_e)P(M)},$$

[2] This reasoning was discussed by the sociologist R. Boudon and would "justify" the error in psychological terms. The disastrous result of the experiment with doctors inspired Tooby and Cosmides to suggest that evolution may have wired the human brain in a "faulty" way regarding statistical intuition.

[3] The reader can verify that even if $P(n|M) \neq P_e$ but still $P(n|M) \ll 1$, the final result does not change.

the formula can be simplified in the case where (as is the case) both $P(M)$ and P_e are small compared to 1:

$$P(M|p) \simeq \frac{1}{1 + P_e/P(M)}. \tag{2.7}$$

With the numerical values of the problem, the probability sought is about 2%.

From (2.7) it is clear that for a laboratory test, the important thing is not so much the error probability of the test P_e as the ratio $P_e/P(M)$: the rarer a disease is, the more accurate the test must be, otherwise the result is not meaningful.

It is not difficult to arrive at the correct answer even without using the formalism. Out of 100,000 subjects, about 100 are sick and 99,900 are healthy. Since the test is wrong in 5% of cases, there will be about 4995 healthy subjects who test positive and about 95 sick people who test positive. Therefore, the probability of being sick given a positive result is about $95/(95 + 4995) \simeq 2\%$.

Three Prisoners Problem

Three men A, B, and C are in prison. Prisoner A has learned that two of them will be executed and one will be freed, but does not know who. The judge has decided "at random" who will be pardoned, so the probability that A will be freed is $1/3$. To the guard, who knows the name of the pardoned man, A says *since two of us will be executed, certainly at least one will be* B *or* C. *You will not tell me anything about my fate, but tell me which of* B *or* C *will be executed.* The guard agrees and says that B *will be executed.* Prisoner A feels a bit more relieved, thinking that either C will be executed or he himself, and concludes that his probability of being freed has risen from $1/3$ to $1/2$. Is he right to be optimistic?

Let $P(A)$ be the probability that A will be freed and $P(b)$ the probability that the guard says that B will be executed. By Bayes' formula, the probability $P(A|b)$ that A will be freed, knowing that B will be executed, is given by

$$P(A|b) = \frac{P(A \cap b)}{P(b)} = P(b|A)\frac{P(A)}{P(b)},$$

where $P(b|A)$ is the probability that B will be executed knowing that A will be freed and, by the law of total probability, $P(b) = P(b|A)P(A) + P(b|B)P(B) + P(b|C)P(C)$. Obviously $P(A) = P(B) = P(C) = 1/3$ while $P(b|A) = 1/2$ (in fact, if A will be freed, the guard will say B or C with equal probability), and also (these need no comment) $P(b|B) = 0$ and $P(b|C) = 1$, we obtain

$$P(b) = \frac{1}{2} \times \frac{1}{3} + 0 \times \frac{1}{3} + 1 \times \frac{1}{3} = \frac{1}{2}$$

and therefore

$$P(A|b) = \frac{\frac{1}{2} \times \frac{1}{3}}{\frac{1}{2}} = \frac{1}{3}.$$

Goats and Cars

The Monty Hall problem originates from the TV game show *Let's Make a Deal*. The contestant is shown three closed doors hiding the prize: behind one is a car and behind the other two, a goat.

After the contestant has chosen a door, but before opening it, the host (who knows what is behind each door) opens another door, revealing one of the two goats, and offers the contestant the chance to switch their initial choice to the only remaining unopened door. Does switching doors improve the contestant's chances of winning the car? The answer is yes: the probability of winning increases from $1/3$ to $2/3$.

Let us consider, without loss of generality, the case in which door 3 is chosen, but no door has yet been opened.

The probability that the car is behind door 2, which we denote as $P(A_2)$, is obviously $1/3$. The probability that the host opens door 1, $P(C_1)$, is $1/2$, since the car has the same probability of being behind door 1 (which would force the host to open door 2) as behind door 2 (which would force the host to open door 1); if the car is not behind either of the two doors (1 or 2), we can assume the host opens one at random, with equal probability. Note that if the car is behind door 2, under these assumptions the host will certainly open door 1, that is, $P(C_1|A_2) = 1$. Using Bayes' theorem, we have:

$$P(A_2|C_1) = \frac{P(C_1|A_2)P(A_2)}{P(C_1)} = \frac{1 \times \frac{1}{3}}{\frac{1}{2}} = \frac{2}{3}$$

so it is advantageous to switch doors.

2.2 Generating Functions: How to Count Without Making Mistakes

Many probability problems involving integer variables can be reduced to combinatorial calculations. Let us consider, for example, the roll of 3 loaded dice such that for the first die, the outcomes $1, 2, \ldots, 6$ occur with probabilities $p_1, p_2, \ldots, p_6$, for the second die with probabilities $q_1, q_2, \ldots, q_6$, and for the third die with probabilities $t_1, t_2, \ldots, t_6$ (obviously with the constraints $\sum_i p_i = \sum_i q_i = \sum_i t_i = 1$), and we are asked for the probability that the sum is 12 or 8. An explicit calculation based only on elementary definitions is clearly possible, but things get more complicated as the number of dice increases.

Fortunately, there is a simple and powerful technique that allows us, so to speak, to *count without explicitly enumerating all possible cases*: the **generating function.**

Consider a random variable x that takes integer values $0, 1, \ldots, k, \ldots$ with probabilities $P_0, P_1, \ldots, P_k, \ldots$, the generating function $G(s)$ is defined as:

$$G(s) = \sum_{k=0}^{\infty} s^k P_k = P_0 + s P_1 + s^2 P_2 + \ldots. = E(s^x). \tag{2.8}$$

The following properties are evident:

$$G(1) = 1 \ , \ G(0) = P_0 \ , \ G'(0) = P_1, \ \ldots, \quad \frac{1}{n!} \frac{d^n G(s)}{ds^n}|_{s=0} = P_n. \tag{2.9}$$

If $x_1, \ldots, x_N$ are independent variables with generating functions $G_1(s), \ldots, G_N(s)$, then the generating function $G_z(s)$ of the sum variable $z = x_1+x_2+\ldots+x_N$ is:

$$G_z(s) = \prod_{i=1}^{N} G_i(s) \ , \tag{2.10}$$

the proof is immediate: it follows from definition (2.8) and from the independence of the variables $x_1, \ldots, x_N$. Formula (2.10) allows us to solve without difficulty the problem of the 3 loaded dice:

$$G_z(s) = (sp_1+s^2p_2+\ldots+s^6p_6)(sq_1+s^2q_2+\ldots+s^6q_6)(st_1+s^2t_2+\ldots+s^6t_6)$$

the probability that the result is k (with $k = 3, 4, \ldots, 18$) is simply the coefficient in front of s^k in $G_z(s)$, a calculation that presents no difficulty.

If generating functions were only useful for problems with loaded dice, or similar things, it would not be so interesting. The underlying idea of the generating function, common to other areas of mathematics, is a kind of "change of basis" (very similar in fact to Fourier transforms), as the knowledge of $G(z)$ is equivalent[4] Computing the $\{P_k\}$, it is sometimes easier, and therefore preferable, to work with the generating functions and then return to the $\{P_k\}$.

An easy but interesting result is the following: if $x_1, \ldots, x_N$ are independent Poisson variables with parameters $\lambda_1, \ldots, \lambda_N$, that is, with probability

$$P(x_i = k) = \frac{\lambda_i^k}{k!} e^{-\lambda_i} \quad k = 0, 1, \ldots.$$

[4] If k takes only finite values, the equivalence $\{P_k\} \leftrightarrow G(s)$ is obvious. In the case $k = 0, 1, \ldots, \infty$ the proof is left to the reader. Hint: note that $G(s)$ is the real part of an analytic function on the real segment $0 \leq s \leq 1$.

then the variable $z = x_1 + x_2 + \ldots + x_N$ is Poisson with parameter $\Lambda = \sum_{i=1}^{N} \lambda_i$. It is enough to compute $G_i(s)$:

$$G_i(s) = \sum_k s^k P_k = \sum_k s^k \frac{\lambda_i^k}{k!} e^{-\lambda_i} = e^{-\lambda_i(1-s)}$$

using (2.10) we have

$$G_z(s) = \prod_{i=1}^{N} e^{-\lambda_i(1-s)} = e^{-\Lambda(1-s)}$$

that is, the generating function of the Poisson distribution with parameter $\Lambda = \sum_{i=1}^{N} \lambda_i$.

2.2.1 Generating Functions and Branching Processes

Let us briefly discuss the use of generating functions for branching processes. In these processes, which are very common in physics and biology, an individual after a certain time (for convenience, taken as unit time) generates k individuals with probability P_k with $k = 0, 1, \ldots$. Branching processes are the probabilistic formalization of chain reactions. For example, think of a neutron that induces a fission that releases other neutrons, which in turn produce further fissions; or organisms that multiply into a certain number of individuals who in turn can multiply.

A rather natural question is: if at time $t = 0$ there is one individual (and the $\{P_k\}$ are known), what is the probability of having m individuals after n generations? If n is small (say 2 or 3) it is not difficult to solve the problem (not elegantly) simply by careful counting; as soon as n increases, the brute force method becomes rather inconvenient.

Fortunately, there is a method to calculate $G_2(s)$, that is, the generating function after 2 generations, and then $G_3(s)$, $G_4(s)$, and so on: the probability of having m individuals after n generations is nothing but the coefficient in front of s^m in $G_n(s)$.

To do this, we must first prove a very useful result: Let $x_1, x_2, x_3, \ldots$ be independent identically distributed variables with generating functions $G(s)$ and N a variable, independent of the $\{x_i\}$, with generating function $F(s)$. The generating function $G_z(s)$ of the variable $z = x_1 + x_2 + \ldots + x_N$ is:

$$G_z(s) = F(G(s)) . \tag{2.11}$$

The proof is simple and follows from (2.8) and (2.9)

$$G_z(s) = E(s^z) = E(s^{x_1 + \ldots + x_N}) = E_N(E_x(s^x)^N) = E_N(G(s)^N) = F(G(s)),$$

where $E_x()$ and $E_N()$ indicate the expectation with respect to the variable x and N, respectively.

With the previous result, we can now calculate $G_n(s)$ for the branching process problem. Let N_{n-1} be the number of individuals in the $(n-1)$-th generation, let x_1 be the number of individuals generated by the first individual, x_2 the number of individuals generated by the second, and so on. At the n-th generation, we will have:

$$N_n = x_1 + x_2 + \ldots + x_{N_{n-1}}.$$

At this point, using (2.11) we have

$$G_n(s) = G_{n-1}(G(s)), \tag{2.12}$$

where $G_2(s) = G(G(s))$, $G_3(s) = G_2(G(s)) = G(G(G(s)))$, ...

In general, the calculation of $G_n()$, though mathematically trivial, can lead to expressions that are cumbersome to handle. This is not always the case. Let us mention an interesting example: around 1920 A. Lotka, using demographic data, showed that in the United States at the beginning of the twentieth century a male generated k males with probabilities given by:

$$P_0 = \frac{1-(\alpha+\beta)}{1-\beta}, \quad P_k = \alpha\beta^{k-1}, \quad k = 1, 2, \ldots$$

with $\alpha \simeq 0.2126$ and $\beta \simeq 0.5892$. The generating function is easily calculated

$$G(s) = P_0 + \alpha \sum_{k=1}^{\infty} \frac{(s\beta)^k}{\beta} = P_0 + \frac{\alpha s}{1-\beta s} = \frac{0.4825 - 0.0717s}{1 - 0.5892s},$$

that is, a hyperbola. A simple calculation shows that

$$G_n(s) = \frac{A_n - B_n s}{C_n - D_n s},$$

that is, $G_n(s)$ is always a hyperbola and the coefficients A_n, B_n, C_n and D_n are obtained with an easy recursive formula from A_{n-1}, B_{n-1}, C_{n-1} and D_{n-1}.

Probability of Extinction

The use of generating functions allows us to answer the question about the probability of extinction, that is, that after many generations there are no longer any descendants of the original progenitor. The calculation is reduced to computing $G_n(0)$ for large n; interestingly, it is not necessary to explicitly determine $G_n(s)$. Let us introduce the probability of extinction after n generations, $e_n = G_n(0)$, then

$$e_1 = G_1(0) = G(0) \ , \ e_2 = G_2(0) = G(G(0)) = G(e_1) \ \dots$$

$$e_n = G_n(0) = G(G_{n-1}(0)) = G(e_{n-1}) \ ,$$

so we have an iterative rule.

Note that $G(1) = 1$, $G'(s) \geq 0$, $G''(s) > 0$, and moreover $G'(1) = E(x)$, so $G(s)$ is a convex, monotonic non-decreasing function. Let us consider the two possible cases:

$$a) \ G'(1) \leq 1 \quad b) \ G'(1) > 1.$$

In case (a), see Fig. 2.1, a graphical iteration of the rule $e_{n+1} = G(e_n)$ shows that e_n tends to 1, that is, extinction is certain. In the second case, see Fig. 2.2, since G is convex, necessarily $G(s)$ intersects the bisector at a point s^* determined by the

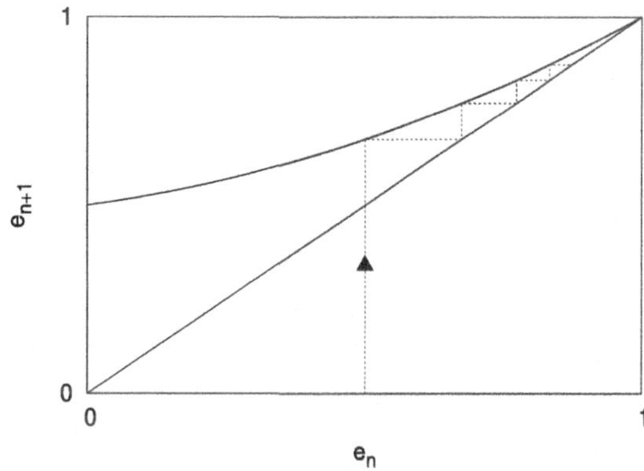

Fig. 2.1 Graphical representation of the iteration of the rule $e_{n+1} = G(e_n)$ in the case where $G'(1) < 1$ and e_n tends to 1

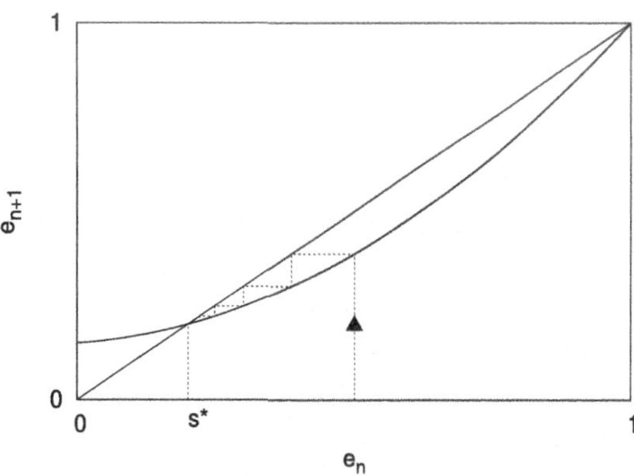

Fig. 2.2 Graphical representation of the iteration of the rule $e_{n+1} = G(e_n)$ in the case where $G'(1) > 1$ and e_n tends to a fixed point s^*, solution of $s^* = G(s^*)$

equation

$$s^* = G(s^*)\,.$$

In the language of dynamical systems, s^* is the fixed point of the map $e_{n+1} = G(e_n)$ and it is easy to see that e_n tends to s^*. Obviously, if $G(0) > 0$ then $s^* > 0$, while if $G(0) = 0$ then $s^* = 0$.

2.3 Some Simple but Useful Results

In this section we discuss some rather simple but useful results, particularly in statistical mechanics.

2.3.1 *How to Change Variables*

Let us consider the case in which we know the probability density $p_x(x)$ of the variable x, and we ask for the probability density $p_y(y)$ of the variable $y = f(x)$. For simplicity, let us start with the case in which $f(x)$ is invertible, that is, $f' \neq 0$. From Fig. 2.3 it is evident that, if $y_1 = f(x_1)$ and $y_2 = f(x_2)$

$$P(y \in [y_1, y_2]) = P(x \in [x_1, x_2]),$$

if $x_2 = x_1 + \Delta x$ with Δx small, then $y_2 = y_1 + \Delta y$ with $\Delta y = f'(x_1)\Delta x$, since

$$p_x(x)\Delta x = p_y(y)|f'(x)|\Delta x\,,$$

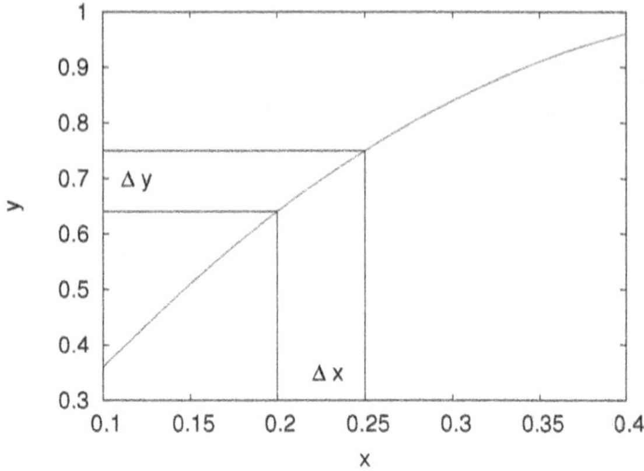

Fig. 2.3 Graphical representation of the change of variables

(the absolute value is introduced to account for cases with $f' < 0$) we obtain

$$p_y(y) = \frac{p_x(x^*)}{|f'(x^*)|} \text{ , with } x^* = f^{-1}(y). \tag{2.13}$$

The case with non-monotonic f is left as an (easy) exercise; we have:

$$p_y(y) = \sum_{x^{(k)}:f(x^{(k)})=y} \frac{p_x(x^{(k)})}{|f'(x^{(k)})|}. \tag{2.14}$$

Equations (2.13) and (2.14) can be written in a compact (and easy to remember) form:

$$p_y(y) = \int p_x(x)\delta(y - f(x))dx \ .$$

In the case of several variables, that is, $y_j = f_j(x_1, \ldots, x_N)$, with $j = 1, .., N$, the procedure is analogous:

$$p_{y_1,\ldots,y_N}(\mathbf{y}) = \sum_{\mathbf{x}^{(k)}:\mathbf{f}(\mathbf{x}^{(k)})=\mathbf{y}} \frac{p_{x_1,\ldots,x_N}(\mathbf{x}^{(k)})}{|det\mathcal{A}(\mathbf{x}^{(k)})|}$$

where $\mathcal{A}$ is the matrix with elements $\partial f_i/\partial x_j$ The previous results, although simple, are conceptually interesting, and there is a connection with Bertrand's paradox discussed in Chap. 1. A straightforward calculation shows that the uniform distribution of the variable x in [0, 1] ($p_x(x) = 1$ if $x \in [0, 1]$) is the probability distribution that maximizes the entropy $-\int p_x(x)\ln[p_x(x)]dx$ with the constraint $p_x(x) = 0$ if x is outside the interval [0, 1], and therefore can be seen as the probability distribution that describes the "maximum uncertainty." From (2.13) it follows that this distribution corresponds to an exponential function for the variable $y = -\ln x$, $p_y(y) = e^{-y}$ if $y \geq 0$ and $p_y(y) = 0$ for $y < 0$, and thus "the uncertainty" seems to decrease. This depends on the fact that the entropy defined as $h = -\int p_x(x)\ln[p_x(x)]dx$ is not an intrinsic quantity, that is, it depends on having used the variable x; it is therefore more appropriate to use the notation $h_x = -\int p_x(x)\ln[p_x(x)]dx$. Let us now calculate $h_y = -\int p_y(y)\ln[p_y(y)]dy$, where $y = f(x)$; from (2.13) we have

$$h_y = -\int p_x(x)[\ln[p_x(x)] - \ln|f'(x)|]dx = h_x + E_x(\ln|f'(x)|) \ .$$

Requiring that h_x be maximal (with certain constraints) leads to an expression for $p_x(x)$; if we now calculate $p_y(y)$ from (2.13) we obtain a different expression from the one that would be obtained by maximizing h_y, with the corresponding constraints. For example, maximizing h_x with the constraint

$$E(x^2) = C^2$$

yields the Gaussian:

$$p_x(x) = \frac{1}{\sqrt{2\pi C^2}} e^{-\frac{x^2}{2C^2}}.$$

Now consider the variable $y = x^3$; from (2.13) we have

$$p_y(y) = \frac{1}{3|y|^{2/3}\sqrt{2\pi C^2}} e^{-\frac{|y|^{2/3}}{2C^2}}.$$

Instead, maximizing h_y with the same constraint

$$E(|y|^{2/3}) = C^2$$

yields a completely different function:

$$p_y(y) = ae^{-b|y|^{2/3}}$$

where the values of a and b depend on C.

In Chap. 9 we will return to the principle of maximum entropy.

2.3.2 *What to Do if Some Variables Are Not of Interest*

We know the joint probability density $p_{x_1,\ldots,x_N}(x_1, \ldots, x_N)$ and we are not interested in all the variables $x_1, \ldots, x_N$ but only in some of them, or in a function $y = f(x_1, \ldots, x_N)$. We will see that these are rather common situations in statistical mechanics. How should we proceed?

Let us begin, for simplicity of notation, with the case of 2 variables: given $p_{x_1,x_2}(x_1, x_2)$, how do we determine $p_{x_1}(x_1)$? The answer is easy:[5]

$$p_{x_1}(x_1) = \int p_{x_1,x_2}(x_1, x_2)dx_2, \tag{2.15}$$

[5] Just note that

$$P(X_1 \in [x_1, x_1 + \Delta x_1]) = \int_{x_1}^{x_1+\Delta x_1} \int_{-\infty}^{\infty} p_{x_1,x_2}(x_1, x_2)dx_1dx_2 = \Delta x_1 \int_{-\infty}^{\infty} p_{x_1,x_2}(x_1, x_2)dx_2.$$

$p_{x_1}(x_1)$ is called the marginal probability density. In the case of 3 variables, one may be interested in only one variable (for example x_1) or two (for example (x_1, x_2)); in this case, we have two types of marginal probability densities:

$$p_{x_1}(x_1) = \int p_{x_1,x_2,x_3}(x_1, x_2, x_3)dx_2dx_3,$$

$$p_{x_1,x_2}(x_1, x_2) = \int p_{x_1,x_2,x_3}(x_1, x_2, x_3)dx_3$$

with obvious generalization.

Let us now discuss the probability density of a function of the random variables: $y = f(x_1, .., x_n)$. In this case as well, the answer is easy:

$$P(y \in [y_1, y_2]) = \int_{y_1 < f(x_1,...,x_n) < y_2} p_{x_1,...,x_n}(x_1, .., x_n)dx_1 \dots dx_N$$

in the limit where y_1 is very close to y_2 we have

$$p_y(y)dy = \int_{y < f(x_1,...,x_n) < y+dy} p_{x_1,...,x_n}(x_1, .., x_n)dx_1 \dots dx_N. \tag{2.16}$$

The previous formula can be written in the easy-to-remember form

$$p_y(y) = \int p_{x_1,...,x_n}(x_1, .., x_n)\delta[y - f(x_1, .., x_n)]dx_1 \dots dx_N,$$

2.4 Applications in Statistical Mechanics

In statistical mechanics, there are several interesting situations in which a projection procedure is used (marginal probability densities), that is, one disregards a class of variables:

(a) in the transition from the microcanonical ensemble to the canonical ensemble;
(b) in the calculation of the probability distribution of energy, or other macroscopic quantities;
(c) in kinetic theory when introducing the one-particle, two-particle, etc., distributions; and in liquid theory where the two-particle distribution plays a fundamental role.

2.4.1 *From the Microcanonical Ensemble to the Canonical Ensemble*

Let us denote by $(\mathbf{X}_1, \mathbf{X}_2)$ the variables that describe the microscopic state of the global system consisting of N particles in a volume V, for which the microcanonical distribution $\rho_M(\mathbf{X}_1, \mathbf{X}_2)$ holds, while $\mathbf{X}_1$ are the microscopic variables of the subsystem (the one that, under appropriate limits, is described by the canonical ensemble). The variables $\mathbf{X}_1$ determine the state of N_1 particles in the volume V_1, similarly $\mathbf{X}_2$ determine the state of the remaining $N_2 = N - N_1$ particles in a volume $V_2 = V - V_1$.

The probability density of the canonical ensemble follows from (2.15):

$$\rho_C(\mathbf{X}_1) = \int \rho_M(\mathbf{X}_1, \mathbf{X}_2) d\mathbf{X}_2 \ .$$

Since in the microcanonical ensemble we have

$$\rho_M(\mathbf{X}_1, \mathbf{X}_2) = \frac{1}{\omega(E, N, V)\Delta} \quad if \quad H(\mathbf{X}_1, \mathbf{X}_2) \in [E, E + \Delta]$$

and $\rho_M(\mathbf{X}_1, \mathbf{X}_2) = 0$ if $H(\mathbf{X}_1, \mathbf{X}_2)$ is outside the interval $[E, E + \Delta]$, with

$$\omega(E, N, V)\Delta = \int_{E<H<E+\Delta} d\mathbf{X}_1 d\mathbf{X}_2.$$

Writing $H(\mathbf{X}_1, \mathbf{X}_2) = H_1(\mathbf{X}_1) + H_2(\mathbf{X}_2) + H_{12}(\mathbf{X}_1, \mathbf{X}_2)$ and neglecting the interaction contribution[6] H_{12} we have

$$\rho_C(\mathbf{X}_1) = \frac{\omega(E - H_1(\mathbf{X}_1), N - N_1, V - V_1)}{\omega(E, N, V)}. \tag{2.17}$$

Recalling that $\omega(E, N, V) = e^{S(E,N,V)/k_B}$ where $S(E, N, V)$ is the microcanonical entropy of the system, k_B is Boltzmann's constant, and $T = \partial E/\partial S$ is the temperature. In the limit $H_1 \ll E$, $N_1 \ll N$ and $V_1 \ll V$, with a Taylor expansion, one obtains the canonical distribution

$$\rho_C(\mathbf{X}_1) = \frac{e^{-\beta H_1((\mathbf{X}_1)}}{Z(T, V_1, N_1)}, \tag{2.18}$$

where $Z(T, V_1, N_1) = \int e^{-\beta H_1((\mathbf{X}_1)} d\mathbf{X}_1$ is the partition function.

[6] This is physically sensible if the interaction range between pairs of particles is small compared to the linear size of the system described by the variables $\mathbf{X}_1$.

2.4.2 *Marginal Probability Densities in Statistical Mechanics*

A very important example of formula (2.16) is the calculation of the probability density of the modulus of velocity in classical statistical mechanics. We know that (Maxwell-Boltzmann distribution)

$$p_{v_x,v_y,v_z}(v_x, v_y, v_z) = P_{MB}(\mathbf{v}) = Be^{-A(v_x^2+v_y^2+v_z^2)}, \tag{2.19}$$

where $A = m/(2k_BT)$ and $B = [m/(2\pi k_BT)]^{3/2}$; for the modulus $v = \sqrt{v_x^2 + v_y^2 + v_z^2}$ from (2.16) we have

$$p_v(v) = 4\pi Bv^2e^{-Av^2} .$$

In the case where $f = E$ (energy), from (2.16) we have

$$p_E(E)dE = const. \int_{E<H<E+dE} e^{-\beta H(\mathbf{X})}d\mathbf{X},$$

where $\mathbf{X}$ denotes the variables that describe the system with Hamiltonian $H(\mathbf{X})$, we can rewrite the previous equation in the form

$$p_E(E)dE = const. e^{-\beta E} \int_{E<H<E+dE} d\mathbf{X} = const. e^{-\beta E}\omega(E)dE,$$

from which, recalling that $\omega(E) = e^{S(E)/k_B}$ one finally obtains

$$p_E(E) = const. e^{-\beta[E-TS(E)]}.$$

It is interesting to discuss the previous result in the limit $N \gg 1$ in which it is possible to assume (and under appropriate conditions explicitly prove):

$$E = eN + o(N), \quad S(E) = s(e)N + o(N)$$

where e and $s(e)$ are the energy per particle and the entropy per particle, respectively. We thus have

$$p_e(e) = const. e^{-\beta[e-Ts(e)]N}.$$

In the next chapter we will see that this type of probability density is a generalization of the central limit theorem (large deviations).

From the Microcanonical Ensemble to the Maxwell-Boltzmann Distribution

It is instructive to derive the Maxwell-Boltzmann distribution using (2.15). Let us consider a system consisting of N non-interacting particles with Hamiltonian

$$H = \sum_{n=1}^{N} \frac{\mathbf{p}_n^2}{2m},$$

in a manner analogous to the procedure used to determine (2.18), for the probability density of the momentum of a particle we have:

$$p_{\mathbf{p}}(\mathbf{p}) = \frac{\omega(E - \frac{\mathbf{p}^2}{2m}, N - 1)}{\omega(E, N)}.$$

Since

$$\omega(E, N) = 3NmC_N(2mE)^{3N/2-1},$$

where $C_N = \pi^{3N/2}/\Gamma(3N/2 + 1)$, in the limit $N \gg 1$

$$p_{\mathbf{p}}(\mathbf{p}) \simeq \frac{C_{N-1}}{C_N} \frac{1}{(2mE)^{3/2}} \left(1 - \frac{\mathbf{p}^2}{2mE}\right)^{3N/2},$$

which is valid for $|\mathbf{p}| \leq \sqrt{2mE}$. Using Stirling's approximation $\Gamma(n+1) \simeq n^n e^{-n}\sqrt{2\pi n}$ we have $C_{N-1}/C_N \simeq (3N/2\pi)^{3/2}$, and recalling that $E = \frac{3}{2}Nk_BT$ we obtain

$$p_{\mathbf{p}}(\mathbf{p}) \simeq \frac{1}{\sqrt{(2\pi mk_BT)^3}}\, e^{-\frac{\mathbf{p}^2}{2mk_BT}}, \tag{2.20}$$

which is nothing but (2.19) written for the momentum instead of the velocity.

Reduced Probability Densities in Kinetic Theory

Let us consider a system of N particles of mass m and denote by $\mathbf{x}_j$ the (6-dimensional) vector that determines the position and momentum of the j-th particle, that is, $\mathbf{x}_j = (\mathbf{q}_j, \mathbf{p}_j)$. The "complete" information about the statistical properties at time t is given by the probability density $\rho_N(\mathbf{x}_1, \ldots, \mathbf{x}_n, t)$ whose evolution is determined by the Liouville equation

$$\frac{\partial \rho_N}{\partial t} + \sum_{j=1}^{N} \frac{\partial \rho_N}{\partial \mathbf{q}_j}\frac{\partial H}{\partial \mathbf{p}_j} - \sum_{j=1}^{N} \frac{\partial \rho_N}{\partial \mathbf{p}_j}\frac{\partial H}{\partial \mathbf{q}_j} = 0,$$

where H is the Hamiltonian of the system. In many circumstances, knowledge of the reduced probability densities is sufficient:

$$\rho_1(\mathbf{x}_1, t) = \int \rho_N(\mathbf{x}_1, \ldots, \mathbf{x}_N, t) d\mathbf{x}_2 d\mathbf{x}_2 \ldots d\mathbf{x}_N,$$

$$\rho_2(\mathbf{x}_1, \mathbf{x}_2, t) = \int \rho_N(\mathbf{x}_1, \ldots, \mathbf{x}_n, t) d\mathbf{x}_3 d\mathbf{x}_4 \ldots d\mathbf{x}_N \quad ,$$

for example, in the Boltzmann equation, which under appropriate assumptions accurately describes the statistical properties of dilute gases, only ρ_1 appears.

Reduced Probability Densities in Liquid State Physics

If the particles interact via a central two-body potential (i.e., depending only on the distance), then the two-particle reduced density is sufficient to determine the thermodynamic properties of the system. In the presence of thermodynamic equilibrium, ρ_2 does not depend on time and has the form

$$\rho_2(\mathbf{x}_1, \mathbf{x}_2) = P_{MB}(\mathbf{p}_1) P_{MB}(\mathbf{p}_2) F_2(\mathbf{q}_1, \mathbf{q}_2)$$

where P_{MB} denotes the Maxwell-Boltzmann probability density (2.20) and $F_2(\mathbf{q}_1, \mathbf{q}_2)$ is the spatial part which, due to the radial symmetry of the problem, will have the form $4\pi r^2 g_2(r)/V$ where $g_2(r)$ is the radial distribution function defined as follows: $4\pi r^2 \rho_0 g_2(r) dr$ is the probability of finding a particle between r and $r + dr$ from a given particle, and $\rho_0 = N/V$ is the particle density of the system. Knowledge of $g_2(r)$ is sufficient to determine the free energy:

$$\frac{U}{N} = \frac{3}{2} k_B T + \frac{\rho_0}{2} \int_0^\infty 4\pi r^2 V_I(r) g_2(r) dr \ , \tag{2.21}$$

where V_I is the interaction potential.

Similarly, from the virial equation for the pressure P one can write the equation of state:

$$P = \rho_0 k_B T - \frac{2\pi \rho_0^2}{3} \int_0^\infty r^3 V_I'(r) g_2(r) dr. \tag{2.22}$$

Equations (2.21) and (2.22) are formally exact, even though it is not simple to calculate $g_2(r)$. However, it is interesting that $g_2(r)$ can be measured with neutron (or light) scattering experiments; moreover, in the dilute gas limit, analytical approximations for $g_2(r)$ can be obtained.

Exercises

2.1 A fair die is rolled 2 times, calculate the probability

(a) of getting a single 6;
(b) that both numbers are even;
(c) that the sum of the numbers is 4;
(d) that the sum of the numbers is a multiple of 3.

2.2 A fair coin is tossed N times, calculate the probability that

(a) the result is heads for the first time on the N-th toss;
(b) heads and tails appear the same number of times (obviously N must be even);
(c) heads comes up exactly 2 times;
(d) heads comes up at least 2 times.

2.3 Industries I and II produce defective items with probabilities $1/5$ and $1/50$ respectively. Knowing that the production of industry I is double that of II, calculate:

(a) the probability of finding a defective item;
(b) the probability that a defective item found comes from industry I.

2.4 Given two independent variables X and Y that take values $k = 1, 2, \ldots$ with probability 2^{-k}, calculate the probability that

(a) $X = Y$;
(b) $X < Y$;
(c) X is three times Y.

2.5 Assume that the probability of a male birth, independently of previous births, is $P = 1/2$ and the following family planning strategy is adopted: keep having children until a boy is born, then stop. Show that, in the limit of infinite lifetime, the average number of boys is equal to that of girls.

2.6 Let $x_1, \ldots, x_N$ be i.i.d. variables with probability density $p_X(x)$, calculate the probability density of the variables

$$y_N = \max\{x_1, \ldots, x_N\}, \quad z_N = \min\{x_1, \ldots, x_N\}.$$

2.7 Let x_1 and x_2 be i.i.d. variables, with Gaussian distribution of zero mean and unit variance, calculate the probability density of the variables

$$r = \sqrt{x_1^2 + x_2^2}, \quad z = \frac{x_2}{x_1}.$$

2.8 Let x_1 and x_2 be i.i.d. variables, with Gaussian distribution, show that the variables

$$z = x_1 + x_2, \; q = x_1 - x_2$$

are Gaussian and independent.

2.9 Let x_1 and x_2 be i.i.d. variables uniformly distributed in [0, 1], consider the variables

$$z_1 = r \cos \psi, \; z_2 = r \sin \psi$$

where

$$r = \sqrt{-2 \ln x_1}, \; \psi = 2\pi x_2.$$

Show that z_1 and z_2 are independent Gaussian variables with zero mean and unit variance.

2.10 Two friends make a probabilistic appointment: they meet at the bar between 17 and 18, wait 5 minutes, and then leave. Assuming that the arrival times are independent and uniformly distributed between 17 and 18, calculate the probability that they meet.

2.11 A random variable takes values $k = 0, 1, \ldots$ with probability P_k. Show that knowing the probabilities $\{P_k\}$ is always equivalent (even in the case where k can take an infinite number of values) to knowing the generating function $G(s) = P_0 + sP_1 + s^2 P_2 + \ldots$, that is, from $G(s)$ one can always determine the probabilities $\{P_k\}$.

2.12 Let X and Y be independent Poisson variables, with parameters λ_x and λ_y respectively, calculate $P(X = k|X + Y = n)$.

2.13 Consider a fair die, calculate the average number of rolls needed to get a 6 for the first time.

2.14 To boost sales, a supermarket puts a sticker numbered from 1 to 5 in each box of cookies. After collecting all the numbers, you are entitled to a free box. Calculate the average number of boxes to buy to get a free one.

Generalize to the case with N stickers.

2.15 Let X and Y be independent with probability density $p_X(x)$ and $p_Y(y)$ respectively, find the probability density of the variables $Z = XY$ and $Q = X/Y$.

2.16 Independence implies uncorrelation, but the converse is not true. Consider the case with

$$P(X = \pm 1, Y = 1) = P(X = \pm 1, Y = -1) = \frac{1}{6},$$

$$P(X = 0, Y = \pm 1) = P(X = \pm 1, Y = 0) = 0, \quad P(X = 0, Y = 0) = \frac{1}{3}.$$

Show that

$$\langle XY \rangle = \langle X \rangle \langle Y \rangle$$

$$P(X = i, Y = j) \neq P(X = i) P(Y = j).$$

Recommended Readings

A useful book to get started:

P. Contucci, S. Isola, *Elementary Probability* (Zanichelli, Bologna, 2008)

For a discussion on the psychological motivations behind errors in probability:

R. Boudon, *Beyond Relativism* (Routledge, London, 2002)

An introduction to probability particularly suitable for physics students:

B.V. Gnedenko, *The Theory of Probability* (MIR Ed., Moscow, 1976)

Two statistical mechanics books with a good discussion on probability:

L.E. Reichl, *A Modern Course in Statistical Physics* (Wiley, New York, 1998);

L. Peliti, *Statistical Mechanics in a Nutshell* (Princeton University Press, Princeton, 2024)

Chapter 3
Limit Theorems: The Statistical Behavior of Systems with Many Variables

3.1 The Law of Large Numbers

Historically, the first example of a limit theorem was the law of large numbers, which, first derived by J. Bernoulli in 1713, forms the basis of the frequentist interpretation of probability. A simple proof can be obtained from the **Chebyshev inequality**:[1]

$$P(|x- < x > | > \epsilon) \leq \frac{\sigma^2}{\epsilon^2}. \tag{3.1}$$

The proof of the previous formula is straightforward:

$$P(|x- < x > | > \epsilon) = \int_{-\infty}^{<x>-\epsilon} p_X(x)dx + \int_{<x>+\epsilon}^{\infty} p_X(x)dx$$

note that in the intervals over which the integral is taken, $|x- < x > | > \epsilon$ so

$$P(|x- < x > | > \epsilon) \leq \int_{-\infty}^{<x>-\epsilon} \frac{(x- < x >)^2}{\epsilon^2} p_X(x)dx +$$

$$+ \int_{<x>+\epsilon}^{\infty} \frac{(x- < x >)^2}{\epsilon^2} p_X(x)dx \leq \int_{-\infty}^{\infty} \frac{(x- < x >)^2}{\epsilon^2} p_X(x)dx = \frac{\sigma^2}{\epsilon^2}.$$

[1] In the theory of probability, Russian names are frequently encountered; since there is no universally accepted rule for the transcription from the Cyrillic to the Latin alphabet, it is easy to find the same name written in different ways, for example Chebyshev is sometimes also written as Tchebichev, similarly Markov and Markoff are the same person, the same goes for Kolmogorov and Kolmogoroff, as well as for Lyapunov, Ljapunov, or Liapounoff. In this book we will follow the transcription used in the Anglo-Saxon literature, which in any case is not without exceptions.

G. Boffetta, A. Vulpiani, *Probability in Physics*, UNITEXT for Physics,
https://doi.org/10.1007/978-3-032-10407-6_3

Similarly, one obtains the **Markov inequality**: for any $k > 0$ we have

$$P(|x - <x>| > \epsilon) \leq \frac{E(|x - <x>|^k)}{\epsilon^k}.$$

Now consider N independent and identically distributed (i.i.d.) variables $x_1, \ldots, x_N$, with mean $<x>$ and variance $\sigma_x^2 < \infty$. The variable $y_N = (x_1 + x_2 + \ldots + x_N)/N$ has mean $<x>$ and variance $\sigma_{y_N}^2 = \sigma_x^2/N$. Now we use the inequality (3.1) for y_N:

$$P\left(\left|\frac{1}{N}\sum_{n=1}^{N} x_n - <x>\right| > \epsilon\right) \leq \frac{\sigma_{y_N}^2}{\epsilon^2} = \frac{\sigma_x^2}{N\epsilon^2}. \tag{3.2}$$

from which it follows that for any $\epsilon > 0$

$$\lim_{N\to\infty} P\left(\left|\frac{1}{N}\sum_{n=1}^{N} x_n - <x>\right| > \epsilon\right) = 0. \tag{3.3}$$

We can therefore say that for $N \to \infty$ the "empirical" mean $\sum_{n=1}^{N} x_n/N$ "converges" to the mean value $<x>$.

The previous result also holds for variables that are not identically distributed, as long as they are independent and have bounded variance: $\sigma_j^2 < B < \infty$. Denoting by m_j the mean of the variable x_j, it is easy to obtain:

$$P\left(\left|\frac{1}{N}\sum_{n=1}^{N}(x_n - m_n)\right| > \epsilon\right) \leq \frac{1}{N^2\epsilon^2}\sum_{n=1}^{N}\sigma_n^2 \leq \frac{B}{N\epsilon^2}.$$

3.1.1 *Something Better than Chebyshev: The Chernoff Inequality*

The estimate (3.2) is rather "generous"; in reality, much more accurate results can be obtained. As an example, consider the case of dichotomous variables: x_n takes the value ± 1 with probability 1/2, so $<x>= 0$ and $\sigma_x^2 = 1$. The following inequality (Chernoff's) holds:

$$P\left(\sum_{n=1}^{N} x_n > a\sqrt{N}\right) \leq e^{-a^2/2}, \tag{3.4}$$

from which it follows

$$P\Big(\Big|\frac{1}{N}\sum_{n=1}^{N} x_n\Big| > \frac{a}{\sqrt{N}}\Big) \leq 2e^{-a^2/2}, \tag{3.5}$$

which is much stronger than (3.2) where $\epsilon = a/\sqrt{N}$:

$$P\Big(\Big|\frac{1}{N}\sum_{n=1}^{N} x_n\Big| > \frac{a}{\sqrt{N}}\Big) \leq \frac{1}{a^2}.$$

Let us prove (3.4). It is easy to obtain[2] the following Markov inequality, which holds for any probability density $p_X(x)$ such that $p_X(x) = 0$ if $x < 0$:

$$P(x \geq b) \leq \frac{< x >}{b}, \tag{3.6}$$

therefore for any non-negative variable x with mean m and variance σ^2 we have

$$P(x \geq m + a\sigma) = P(e^{\lambda x} \geq e^{\lambda(m+a\sigma)}) \leq E(e^{\lambda x})e^{-\lambda(m+a\sigma)}, \tag{3.7}$$

where (3.6) has been used for the variable $e^{\lambda x}$ and λ is an arbitrary constant. Now consider the variable $y_N = x_1 + \ldots + x_N$ where the $\{x_j\}$ are dichotomic variables:

$$E(e^{\lambda y_N}) = E(e^{\lambda x})^N = \Big[\frac{e^{\lambda} + e^{-\lambda}}{2}\Big]^N = (cosh\lambda)^N.$$

Using (3.7) for the variable $y_N = x_1 + \ldots + x_N$, and remembering that $< y_N >= 0$ and $\sigma^2_{y_N} = N$ we obtain:

$$P(y_N \geq a\sqrt{N}) \leq (cosh\lambda)^N e^{-\lambda a\sqrt{N}},$$

since[3] $cosh\lambda \leq e^{\lambda^2/2}$ we have

$$P(y_N \geq a\sqrt{N}) \leq e^{-\lambda a\sqrt{N}+\lambda^2 N/2}.$$

---

[2] Just note that

$$P(x \geq b) = \int_b^{\infty} p_X(x)dx \leq \int_b^{\infty} \frac{x}{b} p_X(x)dx \leq \int_0^{\infty} \frac{x}{b} p_X(x)dx = \frac{< x >}{b}.$$

[3] Let us compare the Taylor expansion of $cosh\lambda$ and $e^{\lambda^2/2}$:

$$cosh\lambda = \sum_{n=0}^{\infty} \frac{\lambda^{2n}}{(2n)!} \qquad e^{\lambda^2/2} = \sum_{n=0}^{\infty} \frac{\lambda^{2n}}{2^n n!}$$

since $(2n)! = (2n)(2n-1)\ldots(n+1)n! \geq 2^n n!$ we have $1/(2n)! \leq 1/(2^n n!)$ and therefore $cosh\lambda \leq e^{\lambda^2/2}$.

The previous inequality holds for any λ, so we can choose it in order to minimize $-\lambda a\sqrt{N} + \lambda^2 N/2$, the optimal value is $\lambda^* = a/\sqrt{N}$ which leads to (3.4). Note that (3.5) holds for any a and N does not need to be large. In the next section we will see that a similar result (but only for $N \gg 1$) holds under very general assumptions.

3.2 Central Limit Theorem

We have seen that in the limit $N \to \infty$ the probability density of $y_n = (x_1 + x_2 + \ldots + x_N)/N$ becomes a Dirac delta centered around $< x >$. A natural question that follows is to ask about the shape of the probability density of the variable $x_1 + x_2 + \ldots + x_N$ in the limit $N \gg 1$ around $N < x >$. We will see that this has a universal form, that is, independent of $p(x)$ of the single variable.

Let us remain in the context of independent variables and ask how to determine the probability density of the sum $z = x + y$ knowing $p_X(x)$ and $p_Y(y)$. A moment of reflection convinces us of the formula

$$p_Z(z) = \int p_X(x) p_Y(y) \delta(z-(x+y)) dx dy = \int p_X(x) p_Y(z-x) dx \equiv (p_X \star p_Y)(z),$$

where $\star$ denotes the convolution. In general, given N independent variables $x_1, \ldots, x_n$ with probability densities $p_1(x_1), \ldots, p_N(x_N)$, for the sum variable $z = x_1 + \ldots + x_N$ we have

$$p_Z(z) = (p_1 \star p_2 \star \ldots \star p_n)(z). \tag{3.8}$$

Apart from a few exceptions, the previous formula is not easy to use.

Among the most important exceptions we mention the case of N Gaussian variables with mean $m_1, m_2, \ldots, m_N$ and variance $\sigma_1^2, \sigma_2^2, \ldots \sigma_N^2$. In fact, using the well-known formula

$$\int_{-\infty}^{\infty} e^{-ax^2+bx} dx = \sqrt{\frac{\pi}{a}} e^{\frac{b^2}{4a}},$$

it is easy to show that z is still a Gaussian variable with mean $m_1 + m_2 + \ldots + m_N$ and variance $\sigma_1^2 + \sigma_2^2 + \ldots + \sigma_N^2$.

Similarly, if $n_1, n_2, \ldots, n_N$ are Poisson variables with parameters $\lambda_1, \lambda_2, \ldots, \lambda_N$:

$$P(n_j = k) = \frac{\lambda_j^k}{k!} e^{-\lambda_j},$$

then $z = n_1 + .. + n_N$ is still a Poisson variable:

$$P(z = k) = \frac{\Lambda^k}{k!} e^{-\Lambda}$$

with $\Lambda = \lambda_1 + \lambda_2 + \ldots + \lambda_N$; this result was already obtained in Sect. 2.2 of Chap. 2, using generating functions.

We therefore need a method to tackle the problem of sums of independent random variables that allows us to control the behavior of (3.8) in the limit $N \gg 1$. The key technical tool is the characteristic function:

$$\phi_x(t) = \int e^{itx} p_X(x) dx = E(e^{itx}). \tag{3.9}$$

Note that, apart from a multiplicative constant, $\phi_x(t)$ is nothing but the Fourier transform of $p_x(x)$. Under fairly general assumptions, the inverse relation with respect to (3.9) holds:

$$p_x(x) = \frac{1}{2\pi} \int \phi_x(t) e^{-itx} dt \ ,$$

and therefore we can say that $\phi_x(t)$ and $p_x(x)$ are equivalent.

Let us show an important property of the characteristic function: if $x_1, x_2, \ldots, x_N$ are independent random variables with characteristic functions $\phi_{x_1}(t)$, $\phi_{x_2}(t), \ldots, \phi_{x_N}(t)$ then for the sum $z = x_1 + x_2 + \ldots + x_N$ we have

$$\phi_z(t) = \prod_{j=1}^{N} \phi_{x_j}(t). \tag{3.10}$$

This result is obtained by noting that for independent variables

$$\phi_z(t) = E(e^{it(x_1+x_2+\ldots+x_N)}) = \prod_{j=1}^{N} E(e^{itx_j}).$$

Another simple (but useful) property of the characteristic function is the following: if the variable x has characteristic function $\phi_x(t)$ then the characteristic function of the variable $y = ax + b$ (where a and b are real constants) is

$$\phi_y(t) = e^{itb} \phi_x(at). \tag{3.11}$$

We are now ready for the central limit theorem (CLT):[4] *asymptotically, the probability density of the sum of many independent variables is a Gaussian.* A bit more precisely: for large N the probability density of

$$z_N = \frac{1}{\sigma\sqrt{N}} \sum_{n=1}^{N} (x_n - < x >)$$

where $x_1, \ldots, x_N$ are i.i.d. variables with mean $< x >$ and variance σ^2, is the Gaussian with zero mean and unit variance:

$$\frac{1}{\sqrt{2\pi}} e^{-z_N^2/2}.$$

To prove the CLT, let us consider the variable $y_N = x'_1 + \ldots + x'_N$ where $x'_j = x_j - < x >$ and let $\phi_{x'}$ denote the characteristic function of x'; from (3.10) we have

$$\phi_{y_N}(t) = [\phi_{x'}(t)]^N,$$

and from (3.11) for the variable $z_N = y_N/(\sigma\sqrt{N})$ we have:

$$\phi_{z_N}(t) = \Big[\phi_{x'}(\frac{t}{\sigma\sqrt{N}})\Big]^N.$$

Note that for small values of t the characteristic function can be written in the form

$$\phi_x(t) = 1 + it < x > -\frac{t^2}{2} < x^2 > +O(t^3),$$

in the case of the variable x' which has zero mean we have

$$\phi_{x'}(t) = 1 - \frac{t^2}{2}\sigma^2 + O(t^3),$$

therefore

$$\phi_{z_N}(t) = \Big[1 - \frac{t^2}{2N} + O(\frac{t^3}{N^{3/2}})\Big]^N$$

[4] The adjective *central* is to be understood as important, fundamental, and refers to the theorem and not to the limit. Perhaps it would be better to say central theorem of the limit.

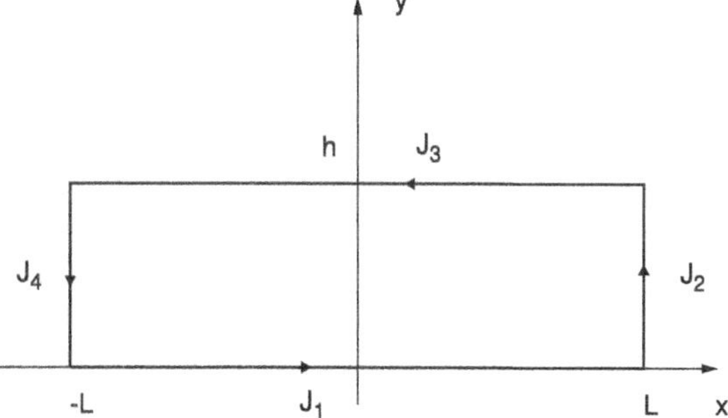

Fig. 3.1 Path for the calculation of the integral of $f(z)$

which in the limit $N \to \infty$ becomes[5]

$$\lim_{N\to\infty} \phi_{z_N} = e^{-\frac{t^2}{2}}.$$

We now show that $\phi(t) = e^{-\frac{t^2}{2}}$ is the characteristic function of the Gaussian with zero mean and unit variance. Let us consider the function of a complex variable

$$f(z) = \frac{1}{\sqrt{2\pi}} e^{-\frac{z^2}{2}+itz}$$

where $z = x + iy$ and t is a real constant. We calculate the integral of $f(z)$ over the closed curve traversed counterclockwise consisting of the rectangle with sides $J_1 : -L < x < L,\ y = 0$; $J_2 : 0 < y < h,\ x = L$; $J_3 : -L < x < L,\ y = h$ and $J_4 : 0 < y < h,\ x = -L$ (see Fig. 3.1). It is easy to see that in the limit $L \to \infty$ the integral over J_2 and J_4 is zero, while the integral over J_1 is nothing but $\phi(t)$, the characteristic function of the Gaussian with zero mean and unit variance; similarly, a straightforward calculation shows that the integral over J_3 is $-\phi(t-h)e^{\frac{h^2}{2}-th}$; since $f(z)$ is analytic, the integral over the closed circuit is zero, thus

$$\phi(t) = \phi(t-h)e^{\frac{h^2}{2}-th},$$

recalling that $\phi(0) = 1$ and setting $h = t$ we obtain the result

$$\phi(t) = e^{-\frac{t^2}{2}}.$$

[5] We are assuming that it is possible to neglect the $O(t^3)$ terms, which is correct if $p_x(x)$ decays sufficiently rapidly for large $|x|$. This is a technical point that will be addressed later.

Therefore, in the limit $N \to \infty$

$$p_{Z_N}(z) \to \frac{1}{\sqrt{2\pi}} e^{-\frac{z^2}{2}}.$$

Note that in the final result the details of $p_x(x)$, apart from $< x >$ and σ, have disappeared.

In a more mathematically precise form we have:

$$\lim_{N\to\infty} P(a < z_N < b) = \frac{1}{\sqrt{2\pi}} \int_a^b e^{-x^2/2} dx. \tag{3.12}$$

The condition that the variables are identically distributed is not essential and can be removed without difficulty, provided they are independent with finite and nonzero variance: $0 < \sigma_j^2 < \infty$. The calculation is repeated in a similar way for

$$z_N = \frac{1}{D_N} \sum_{n=1}^{N} (x_n - m_n),$$

where m_j is the mean value of x_j and

$$D_N^2 = \sum_{n=1}^{N} \sigma_n^2.$$

Let us denote $y_N = (x_1 - m_1) + (x_2 - m_2) + \ldots + (x_N - m_N)$, we have

$$\phi_{y_N}(t) = \prod_{j=1}^{N} \phi_{x_j'}(t) = \prod_{j=1}^{N} \Big[1 - \sigma_j^2 \frac{t^2}{2} + O(t^3)\Big],$$

thus

$$\phi_{z_N}(t) = \phi_{y_N}\Big(\frac{t}{D_N}\Big) = \prod_{j=1}^{N} \Big[1 - \frac{\sigma_j^2}{2} \frac{t^2}{D_N^2} + O(t^3)\Big] =$$

$$= exp \sum_{n=1}^{N} \ln\Big(1 - \frac{\sigma_n^2}{2} \frac{t^2}{D_N^2} + O(t^3)\Big),$$

since $0 < \sigma_j^2 < \infty$ we have that $D_N^2 \sim N$; therefore the coefficient multiplying t^2 is small and we can write

$$\phi_{y_N}(t) \simeq exp\Big[-\sum_{n=0}^{N} \frac{\sigma_n^2}{2} \frac{t^2}{D_N^2}\Big] = e^{-\frac{t^2}{2}}.$$

The mathematically precise hypothesis for the central limit theorem for independent variables is the validity of Lindeberg's condition: for every $\tau > 0$ one must have

$$\lim_{N\to\infty} \frac{1}{D_N^2} \sum_{n=1}^{N} \int_{|x-m_n|>\tau D_n} (x - m_n)^2 p_{x_n}(x)dx = 0,$$

which is equivalent to requiring

$$\lim_{N\to\infty} \frac{1}{D_N^2} \sum_{n=1}^{N} \int_{|x-m_n|<\tau D_n} (x - m_n)^2 p_{x_n}(x)dx = 1.$$

In intuitive terms, Lindeberg's condition means that the single variance σ_n^2 is small compared to D_n^2: for any τ, for large n we have $\sigma_n < \tau D_n$.

The central limit theorem explains[6] the fact that the Gaussian distribution appears in very different and apparently unrelated situations: from physics to biology, from economics to the social sciences. In fact, in many cases the value of a variable is the result of many independent causes.

Let us note that in the proof of the central limit theorem, the two fundamental ingredients that allow for a simple demonstration are:

(a) finite variance $\sigma_x^2 < \infty$;
(b) independence of the variables $\{x_j\}$.

It is not difficult to see that if $\sigma_x^2 = \infty$ the sum of many independent variables does not approach a Gaussian. An example that is easy to treat analytically is the case of independent variables whose probability distribution is:

$$p_X(x) = \frac{1}{\pi(1 + x^2)}.$$

This distribution is called the Cauchy distribution, and with a simple calculation using complex analysis, it can be shown that its characteristic function is $\phi_x(t) = e^{-|t|}$. Let us consider the variable $y_N = x_1 + \ldots + x_N$, where the $\{x_j\}$ are independent and distributed according to the Cauchy function, then $\phi_{y_N}(t) = e^{-N|t|}$ and thus the average y_N/N is distributed like a single x independently of N. Therefore, neither the law of large numbers nor the CLT holds. In Chap. 8 we will return to the general problem of the sum of independent variables with infinite variance.

---

[6] We say explains, and not proves, because in the natural sciences one never has a "true proof"; in fact, it is practically impossible to have complete control over the hypotheses. For example, it is not easy to have empirical certainty of independence. We will see later that there is another distribution (the lognormal) very common in natural phenomena; this distribution has a close connection with the central limit theorem.

3.2.1 What Happens if the Variables Are Not Independent?

As for the problem of non-independent variables, intuitively one expects that if the $\{x_j\}$ are only "weakly dependent" then the central limit theorem still holds and the only modification is to replace σ^2 with an effective variance σ^2_{eff} that takes correlations into account. This problem will be revisited in Chap. 8 when we address the problem of diffusion. We now present a heuristic argument to define the limit of "weakly dependent".

Let us consider the "block" variables[7]

$$Y_k^{(L)} = \frac{1}{\sqrt{L}} \sum_{n=(k-1)L+1}^{kL} x_n ,$$

for simplicity of notation, we consider the case with $\langle x_j \rangle = 0$. Let us calculate the variance of $Y_k^{(L)}$: a straightforward calculation shows

$$\sigma^2_{Y^{(L)}} = E\Big[(Y_k^{(L)})^2\Big] = \sigma^2 + \frac{2}{L}\sum_{n<m} E(x_n x_m),$$

where the indices n and m range between $(k-1)L+1$ and kL. Let us assume that the sequence $x_1, x_2, \ldots$ is statistically stationary[8] and introduce the correlation function $C(k) = E(x_0 x_k) = E(x_n x_{n+k})$. There are two possibilities:

$$a) \sum_{k=1}^{\infty} C(k) < \infty \qquad b) \sum_{k=1}^{\infty} C(k) = \infty. \tag{3.13}$$

In the first case, for $L \gg 1$ we have

$$\sigma^2_{Y^{(L)}} = \sigma^2 + 2\sum_{k=1}^{\infty} C(k) < \infty,$$

in the second case the variance diverges as $L \to \infty$. In case a), the variables $Y_k^{(L)}$ in the limit $L \gg 1$ are practically uncorrelated (which is not exactly independence but is in the right direction); in fact, it is easy to show that

$$E\Big(Y_k^{(L)} Y_{k+1}^{(L)}\Big) = \frac{1}{L}\sum_{l=1}^{L} l C(l).$$

[7] This way of grouping the variables is common in the treatment of critical phenomena and was introduced by L. Kadanoff.

[8] In other words, the "degree of dependence" between x_k and x_l depends only on $k - l$.

If $\sum_{k=1}^{\infty} C(k) < \infty$ then the previous quantity tends to zero[9] for large values of L.

Therefore, if $\sum_{k=1}^{\infty} C(k) < \infty$, for large L the variables $Y_k^{(L)}$ are uncorrelated and we expect the central limit theorem to hold. Writing $N = LM$ we have

$$\frac{1}{\sqrt{N}} \sum_{n=1}^{N} x_n = \frac{1}{\sqrt{M}} \sum_{k=1}^{M} Y_k^{(L)}$$

so if the variables $\{x_j\}$ are weakly correlated, i.e., $\sum_{k=1}^{\infty} C(k) < \infty$, then the central limit theorem still holds with σ^2 replaced by

$$\sigma_{eff}^2 = \sigma^2 + 2 \sum_{k=1}^{\infty} C(k). \tag{3.14}$$

The previous result, which we have argued heuristically, has been proven under general assumptions (essentially the Lindeberg condition and the validity of condition (a) in (3.13)). It is interesting that there exists an important class of stochastic processes (ergodic Markov chains with a finite number of states) whose correlations decay exponentially and for which, therefore, the CLT holds.

3.3 Large Deviations

In the previous section, we saw that the central limit theorem, under suitable assumptions, proves (3.12). Sometimes the CLT is stated by saying that given N independent variables, for the variable $y_N = (x_1 + \ldots + x_N)/N$ when $N \gg 1$ one has

$$p_{y_N}(y_N) \simeq \frac{1}{\sqrt{2\pi\sigma^2/N}} e^{-(y_N - m)^2 N/(2\sigma^2)}, \tag{3.15}$$

this is essentially correct if one adds that $y_N - m$ must not be too large, say no more than $O(\sigma/\sqrt{N})$. It is, however, incorrect to assume the validity of (3.15) literally, that is, over arbitrary intervals. This is evident if we consider the case with $x_n = \pm 1$ with equal probability; since $< x >= 0$ and $\sigma^2 = 1$, in the limit $N \gg 1$ the variable $z_N = \sum_{n=1}^{N} x_n/\sqrt{N}$ has as probability density $e^{-z_N^2/2}/\sqrt{2\pi}$. Note that for $|z_N| > N$ the true probability density is obviously zero, while the Gaussian never vanishes. Moreover, (3.15) is not able to describe certain important statistical

[9] In fact, if $C(l)$ decays asymptotically exponentially, or as $O(l^{-\alpha})$ with $\alpha > 2$ then $\sum_{l=1}^{L} lC(l)$ is finite. If $C(\ell) = O(l^{-\alpha})$ with $1 < \alpha \leq 2$ then $\sum_{l=1}^{L} lC(l) \sim L^{2-\alpha}$ so $E\left(Y_k^{(L)} Y_{k+1}^{(L)}\right) = O(L^{1-\alpha})$ is negligible (since $\alpha > 1$) for $L \gg 1$.

aspects when $y_N - m$ is not small compared to $\sigma/\sqrt{N}$, as becomes clear from the following example. Let us consider $x_1, \ldots, x_N$ i.i.d. and bounded variables $a < x_j < b$. We ask what is the probability density of

$$y_N = \prod_{n=1}^{N} x_n,$$

as we will see, this problem is interesting in various contexts. Taking the logarithm, we have

$$y_N = exp\Big[\frac{1}{N}\sum_{n=1}^{N} \ln x_n\Big]N,$$

since the $\{x_n\}$ are independent, so are $\{\alpha_j = \ln x_j\}$, therefore for the variable $A_N = \ln y_n = \alpha_1 + \ldots + \alpha_N$, invoking the CLT we conclude that

$$p_{A_N}(A_N) \simeq \frac{1}{\sqrt{2\pi C^2 N}} e^{-(A_N - <\alpha> N)^2/(2C^2N)},$$

where C^2 is the variance of α. Using the change of variables rule seen in Chap. 2, for y_N we have:

$$p_{Y_N}(y_N) \simeq p_{LN}(y_N) = \frac{1}{y_N\sqrt{2\pi C^2 N}} e^{-(\ln y_N - <\alpha> N)^2/(2C^2N)}, \tag{3.16}$$

where $p_{LN}(\,)$ is called the lognormal distribution. Taking the previous result literally, that is, assuming the validity of the lognormal even outside the limits of validity of the CLT ($|\delta A_N| < O(C\sqrt{N})$), one obtains manifestly inconsistent results. Let us calculate, for example, $E(y_N^q)$; from (3.16) we have

$$E_{LN}(y_N^q) = e^{N(q<\alpha>+q^2C^2/2)}, \tag{3.17}$$

where LN indicates that the mean value is calculated with the lognormal distribution. The exact result is

$$E(y_N^q) = [E(x^q)]^N = e^{N \ln E(x^q)}$$

since $a < x_n < b$ we have $E(x^q) < b^q$ and therefore

$$E(y_N^q) < e^{Nq \ln b}$$

in clear contrast with (3.17) for sufficiently large q. This is due to the fact that the true probability density of y_N is exactly zero for $y_N > b^N$, while the dominant contribution to the calculation of $E_{LN}(y_N^q)$ for large q comes precisely from values of $y_N > b^N$. In other words, denoting by $p_V(y_N)$ the exact probability distribution, it is true that

$$p_{LN}(y_N) \simeq p_V(y_N),$$

however, it is not correct to state that

$$p_{LN}(y_N) y_N^q \simeq p_V(y_N) y_N^q,$$

for arbitrary values of q. Instead, for small values of q, the lognormal approximation (3.17) is correct. It is enough to note that for q close to zero we have

$$\ln E(x^q) = \ln E(e^{q \ln x}) = \ln E\Big(1 + q \ln x + \frac{q^2}{2}(\ln x)^2 + \ldots\Big),$$

recalling that for small ϵ the expansion $\ln(1 + \epsilon) = \epsilon - \epsilon^2/2 + O(\epsilon^3)$ holds, we have

$$\ln E(x^q) = q < \ln x > + \frac{q^2}{2} < (\ln x - < \ln x >)^2 > + O(q^3)$$

$$= q < \alpha > + C^2 \frac{q^2}{2} + O(q^3) = \ln E_{LN}(x^q) + O(q^3).$$

3.3.1 Beyond the Central Limit: The Cramer Function

From the previous example, it is clear that it is necessary to go beyond the CLT, that is, to master how the "large deviations" behave. This theory was introduced in the 1930s by the Swedish mathematician H. Cramer to describe the statistics of rare events in the context of insurance risks. The idea can be explained with a simple combinatorial calculation.

Let us consider a sequence of tosses of a biased coin whose possible outcomes are heads $(+1)$ or tails (-1), and we denote the result of the n-th toss by x_n, where heads has probability p and tails $1 - p$. Let $y_N = (x_1 + \ldots + x_N)/N$; obviously $\langle y_N \rangle = 2p - 1$ and $\sigma_{y_N}^2 = 4p(1 - p)/N$.

The number of ways in which one can get heads k times in N tosses is $N!/[k!(N - k)!]$, so from the binomial distribution we have

$$P\Big(y_N = \frac{2k}{N} - 1\Big) = \frac{N!}{k!(N - k)!} p^k (1 - p)^{N-k}. \tag{3.18}$$

Using Stirling's approximation $n! \simeq n^n e^{-n}\sqrt{2\pi n}$ and writing $k = fN$ and $N-k = (1-f)N$ where $f = k/N$ is the frequency of the heads event in N tosses, we have

$$P(y = 2f - 1) \sim e^{-NI(p,f)}, \tag{3.19}$$

where

$$I(p, f) = f \ln \frac{f}{p} + (1 - f) \ln \frac{1 - f}{1 - p}. \tag{3.20}$$

The quantity $I(p, f)$ is called "relative entropy" (or Kullback-Leibler divergence); we have $I(p, f) = 0$ if $f = p$, while $I(p, f) > 0$ for $f \neq p$. It is easy to repeat the argument in the multinomial case where $x_1, \ldots, x_N$ can take m possible different values $a_1, a_2, \ldots, a_m$ with probabilities $p_1, p_2, \ldots, p_m$. In the limit $N \gg 1$, the probability of observing the frequencies $f_1, f_2, \ldots, f_m$ is

$$P_N(\{f_j\}) \sim e^{-NI(\{p\},\{f\})}$$

where

$$I(\{p\}, \{f\}) = \sum_{j=1}^{m} f_j \ln \frac{f_j}{p_j},$$

is the relative entropy of the probabilities $\{f\}$ with respect to the probabilities $\{p\}$. This quantity measures the "distance"[10] between $\{p\}$ and $\{f\}$ in the sense that $I(\{p\}, \{f\}) = 0$ if and only if $\{p\} = \{f\}$, and $I(\{p\}, \{f\}) > 0$ if $\{p\} \neq \{f\}$.

From the previous calculation, it is clear how it is possible to go beyond the central limit theorem and control the statistical properties of extreme events (tails of the probability distribution) for $N \gg 1$. Writing $I(f, p)$ in terms of $y = 2f - 1$,

---

[10] In fact, it is not a true distance in the technical sense. Given two vectors $\mathbf{x}$ and $\mathbf{y}$, a function $d(\mathbf{x}, \mathbf{y})$ is a distance if

(a) $d(\mathbf{x}, \mathbf{y})$ is positive except when $\mathbf{x} = \mathbf{y}$ in which case it is zero;
(b) $d(\mathbf{x}, \mathbf{y}) = d(\mathbf{y}, \mathbf{x})$;
(c) $d(\mathbf{x}, \mathbf{z}) \leq d(\mathbf{x}, \mathbf{y}) + d(\mathbf{y}, \mathbf{z})$;

the last inequality (called the triangle inequality) does not hold for relative entropy. Condition (b) does not hold either, but this is not a serious problem; it is enough to symmetrize things and consider

$$\frac{1}{2}[I(\{p\}, \{f\}) + I(\{f\}, \{p\})].$$

Eq. (3.19) becomes

$$p_{y_N}(y_N) \sim e^{-NC(y_N)}, \tag{3.21}$$

with

$$C(y) = \frac{1+y}{2} \ln \frac{1+y}{2p} + \frac{1-y}{2} \ln \frac{1-y}{2(1-p)}.$$

The function $C(y)$ is called the Cramer function. For values of f close to p, that is, $y \simeq \langle y \rangle$, the Taylor expansion shows that

$$C(y) \simeq \frac{(y - \langle y \rangle)^2}{2\sigma^2},$$

with $\sigma^2 = 4p(1-p)$, in agreement with what is expected from the central limit theorem. Equation (3.21) has a general validity (within the context of i.i.d. variables) and can be obtained with a different approach that allows one to express $C(y)$, where $y_N = (x_1 + \ldots + x_N)/N$, in terms of the moments of the variable x. In particular, it is possible to show that the Cramer function $C(y)$ can be written with a Legendre transform:

$$C(y) = \sup_q \Big[qy - L(q) \Big], \tag{3.22}$$

where $L(q)$ is the "cumulant generating function":

$$L(q) = \ln E(e^{qx}). \tag{3.23}$$

Let us briefly outline the argument. Consider the moments $E(e^{qNy_N})$ which can be written in two different ways:

$$E(e^{qNy_N}) = E(e^{qx})^N = e^{NL(q)}$$

$$E(e^{qNy_N}) = \int e^{qNy_N} p_{y_N}(y_N) dy_N \sim \int e^{[qy - C(y)]N} dy, \tag{3.24}$$

in the limit of large N, using the Laplace method one obtains

$$L(q) = \sup_y \Big[qy - C(y) \Big], \tag{3.25}$$

which is the inverse of (3.22). Now, since it can be shown that $C(y)$ is a concave function ($d^2C/dy^2 \geq 0$), Eqs. (3.22) and (3.23) are equivalent.

Let us note that the Cramer function must obey some constraints:

(a) $C(y) > 0$ for $y \neq \langle y \rangle$;
(b) $C(y) = 0$ for $y = \langle y \rangle$;
(c) if y is close to $\langle y \rangle$ then $C(y) \simeq (y - \langle y \rangle)^2/(2\sigma^2)$, where $\sigma^2 = \langle (x - \langle x \rangle)^2 \rangle$;
(d) $C(y)$ is a concave function ($d^2C/dy^2 \geq 0$).

Obviously, (a) and (b) are consequences of the law of large numbers and (c) is nothing but the central limit theorem. (d) is less intuitive; later we will see its meaning in statistical mechanics.

In the more general and interesting case of dependent variables, $L(q)$ is defined by

$$L(q) = \lim_{N \to \infty} \frac{1}{N} \ln E\left(e^{q \sum_{n=1}^{N} x_n}\right),$$

and (3.22) is exact if $C(y)$ is concave, otherwise it provides the concave envelope of $C(y)$.

3.4 Large and Small Fluctuations in Statistical Mechanics

As mentioned at the end of Chap. 2, in statistical mechanics large deviations naturally appear in the problem of energy fluctuations per particle of a system with N particles at temperature T:

$$p(e) \simeq \frac{1}{C_N} exp\{-N\beta[e - Ts(e)]\}, \tag{3.26}$$

where $s(e)$ is the microcanonical entropy density. Since $\int p(e)de = 1$, the constant C_N (partition function) can be expressed as:

$$C_N \sim exp\{-N\beta f(T)\},$$

where $f(T)$ is the free energy per particle

$$f(T) = \min_e \{e - Ts(e)\} \ .$$

The value e^*, for which the function $e - Ts(e)$ is minimum, is determined by the equation

$$\frac{1}{T} = \frac{\partial s(e)}{\partial e}, \tag{3.27}$$

that is, the value of the energy such that the corresponding microcanonical ensemble has temperature T. The Cramer function and its physical meaning are evident:

$$C(e) = \beta[e - Ts(e) - f(T)].$$

Note that the value of e for which $C(e)$ is minimum (zero) is precisely $e^* =< e >$ determined by (3.27). The Gaussian approximation around e^* is

$$C(e) \simeq \frac{1}{2}C''(e^*)(e - e^*)^2,$$

thus $< (e - e^*)^2 >= 1/[NC''(e^*)]$ which is equivalent to

$$< (e - e^*)^2 >= \frac{k_B}{N}T^2 c_V, \tag{3.28}$$

where $c_V = \partial < e > /\partial T$ is the specific heat per particle. The concavity of $C(e)$ has a clear physical counterpart: $c_V(T)$ must be positive so that $< (e - e^*)^2 >$ is positive.

3.4.1 Einstein's Theory of Fluctuations

Equation (3.26) can be generalized to the case of fluctuations of other quantities (besides energy) of thermodynamic interest. Let us assume that the macroscopic state is described by n variables (which also include energy), $\alpha_1, \ldots, \alpha_n$, which are functions of the microscopic state $\mathbf{X}$: $\alpha_j = g_j(\mathbf{X})$, $j = 1, \ldots, n$. We denote by $\mathcal{P}$ the vector of parameters that determine the probability density of the microscopic state $\mathbf{X}$; for example, in the canonical ensemble $\mathcal{P} = (T, N, V)$, in the grand canonical ensemble $\mathcal{P} = (T, \mu, V)$, in the isobaric ensemble at fixed temperature $\mathcal{P} = (T, N, P)$. The probability density of the $\{\alpha_j\}$ is

$$p(\alpha_1, \ldots, \alpha_n) = \int \rho(\mathbf{X}, \mathcal{P}) \prod_{j=1}^{n} \delta(\alpha_j - g_j(\mathbf{X}))d\mathbf{X}$$

where $\rho(\mathbf{X}, \mathcal{P})$ is the probability density of the state $\mathbf{X}$ in the ensemble characterized by the thermodynamic parameters $\mathcal{P}$. Proceeding in a manner analogous to what was done for the probability density of the energy (see Sect. 2.4), we obtain:

$$p(\alpha_1, \ldots, \alpha_n) \sim exp - \beta\Big[F(\alpha_1, \ldots, \alpha_n|\mathcal{P}) - F(\mathcal{P})\Big], \tag{3.29}$$

where $F(\mathcal{P})$ is the free energy of the system with parameters $\mathcal{P}$, while $F(\alpha_1, \ldots, \alpha_n|\mathcal{P})$ is the free energy of the system in which the parameters $\mathcal{P}$

and the macroscopic variables $\alpha_1, \ldots, \alpha_n$ are fixed. For example, in the canonical ensemble where $\mathcal{P} = (T, N, V)$, we have

$$F(\alpha_1, \alpha_2, \ldots, \alpha_n|\mathcal{P}) = -k_B T \ln \int \prod_{j=1}^{n} \delta(\alpha_j - g_j(\mathbf{X}))\, e^{-\beta H(\mathbf{X})} d\mathbf{X} \, .$$

We can identify the quantity

$$\delta S(\alpha_1, \ldots, \alpha_n) = \frac{1}{T}\Big[F(\alpha_1, \ldots, \alpha_n|\mathcal{P}) - F(\mathcal{P})\Big], \tag{3.30}$$

with the difference in entropy between the state with parameters $(\alpha_1, \ldots, \alpha_n, \mathcal{P})$ and that with $\mathcal{P}$. From (3.29) and (3.30) we have:

$$p(\alpha_1, \ldots, \alpha_n) \sim exp \, \frac{\delta S(\alpha_1, \ldots, \alpha_n)}{k_B},$$

this relation is called the Boltzmann–Einstein principle.

In the case where it is possible to show that $F(\alpha_1, \ldots, \alpha_n|\mathcal{P})$ and $F(\mathcal{P})$ are extensive quantities, that is, for large N,

$$F(\alpha_1, \ldots, \alpha_n|\mathcal{P}) \simeq N f\left(\frac{\alpha_1}{N}, \ldots, \frac{\alpha_n}{N}|\mathcal{P}\right),$$

the problem can be described in terms of large deviations.

In macroscopic systems, it is expected, and analytical calculations confirm, that fluctuations with respect to the thermodynamic equilibrium state are (for extensive variables) small in percentage. It is therefore justified to expand $\delta S(\alpha_1, \alpha_2, \ldots, \alpha_n)$ in a Taylor series around the mean values of $\{\alpha_j\}$, which are precisely the values at thermodynamic equilibrium $\{\alpha_j^*\}$:

$$\delta S(\alpha_1, \ldots, \alpha_n) \simeq -\frac{1}{2} \sum_{i,j} \delta\alpha_i A_{ij} \delta\alpha_j \tag{3.31}$$

where

$$\delta\alpha_j = \alpha_j - \alpha_j^*, \quad A_{ij} = -\frac{\partial S}{\partial\alpha_j \partial\alpha_i}\Big|_{\alpha^*}.$$

We thus have that small fluctuations around the thermodynamic equilibrium values are described by a multivariate Gaussian distribution:

$$p(\alpha_1, \ldots, \alpha_n) \simeq \sqrt{\frac{det\mathbf{A}}{(2\pi k_B)^N}} \, exp - \frac{1}{2k_B} \sum_{i,j} \delta\alpha_i A_{ij} \delta\alpha_j,$$

from which

$$\langle \delta\alpha_i \delta\alpha_j \rangle = k_B \Big[\mathbf{A}^{-1} \Big]_{ij} . \tag{3.32}$$

The previous relations are the basis of the so-called Einstein fluctuation theory. Note that the matrix elements A_{ij} are calculated at thermodynamic equilibrium. The matrix **A** must be positive definite (i.e., with strictly positive eigenvalues); this property has a precise physical meaning: the difference in entropy (with respect to thermodynamic equilibrium) must be negative.

Equation (3.28) is a particular case of (3.32); another important example is the fluctuations of the magnetization per particle:

$$\langle (\delta m)^2 \rangle = \frac{k_B}{N} T \chi \tag{3.33}$$

where χ is the magnetic susceptibility per particle. It is interesting that the variance of the fluctuation of a given quantity is proportional to an appropriate response function. In (3.29) the specific heat $c_V = \partial \langle e \rangle / \partial T$ appears, which measures how the average energy changes as the temperature varies, while in (3.33) the susceptibility $\chi = \partial \langle m \rangle / \partial B$ comes into play, which determines how the average magnetization changes with the magnetic field. This is a general result: the *fluctuation-dissipation theorem*.

Note, see (3.28) and (3.32), that the relative fluctuations are small for $N \gg 1$. This might suggest that in macroscopic systems fluctuations are irrelevant and not measurable. On the contrary, they are conceptually very important: as noted by Einstein: *if it were possible to measure the energy fluctuations of the system, one would have an exact determination of the fundamental constant* k_B and thus of Avogadro's number; this aspect will be discussed in detail when we will deal with Brownian motion. The fluctuation-response theorem, of which (2.27) and (2.32) are particular cases, implies that fluctuations are (indirectly) measurable since they are connected to response functions (such as specific heat and susceptibility).

3.5 Some Applications of Limit Theorems Beyond Physics

We conclude the chapter on limit theorems with a brief discussion of their applications in non-physical contexts. A digression that can be seen both as an amusement, and as an introduction to other fields.

3.5.1 Law of Large Numbers and Finance

Let us briefly discuss the use of the law of large numbers in investment strategy in a market that includes banks and stocks. The original idea seems to date back to D. Bernoulli (early eighteenth century) and was developed starting from the 1950s, in particular for its connection with information theory. Let us begin with the case of a very simple market in which there are only two possibilities:

(a) put the money in the bank;
(b) invest it in the stock market, which consists of a single stock.

Let us consider the time at which financial operations can be carried out as discrete (say, in days). In the bank there are no surprises: 1 Euro today remains 1 Euro tomorrow (for simplicity, we consider the case of zero interest). In the stock market, however, the stock price varies randomly; let p_t be the price of the stock at time t, and by investing 1 Euro at time t, one will have $u_t = p_{t+1}/p_t$ Euros at time $t + 1$. We assume that the $\{u_t\}$ are independent random variables[11] with known probability density $p_u(u)$.

Obviously, we can decide to invest a fraction ℓ of the capital in the stock market and the remaining part in the bank. Letting S_t denote the capital at time t, we have

$$S_{t+1} = (1 - \ell)S_t + u_t \ell S_t = X_t(\ell)S_t, \tag{3.34}$$

where we have introduced the random variable $X_t(\ell) = [1+(u_t-1)\ell]$. The question is how to determine ℓ so as to maximize the long-term gain. From (3.34) we have

$$\frac{S_T}{S_0} = \prod_{t=1}^{T} X_t(\ell) = exp\Big[\frac{1}{T}\sum_{t=1}^{T} \ln X_t(\ell)\Big]T;$$

by the law of large numbers, the quantity

$$\frac{1}{T}\sum_{t=1}^{T} \ln X_t(\ell)$$

in the limit $T \to \infty$ is "close" to $\langle \ln X(\ell)\rangle = \int \ln[1 + (u - 1)\ell]p_u(u)du$, with probability close to one (see (3.2)).

At this point, the strategy seems clear, at least if one is not in a hurry and is interested in long time scales:[12] one must set ℓ so as to maximize $< \ln X(\ell) >$. Let us consider the case in which u takes only discrete values $u^{(1)}, u^{(2)}, \dots, u^{(M)}$ with

[11] The generalization to the Markovian case does not present particular difficulties.

[12] It could (rightly) be noted that over long time scales the $\{u_t\}$ are not stationary since the market changes qualitatively.

probabilities $p_1, p_2, \dots, p_M$ so that

$$< \ln X(\ell) >= \sum_j p_j \ln[1 + (u^{(j)} - 1)\ell],$$

and the optimal value ℓ_* is given by the solution of the equation

$$\sum_j p_j \frac{(u^{(j)} - 1)}{1 + (u^{(j)} - 1)\ell_*} = 0.$$

Introducing the quantities $q_j(\ell_*) = p_j/[1 + (u^{(j)} - 1)\ell_*]$ we have

$$\langle \ln X(\ell_*) \rangle = \sum_j p_j \ln \frac{p_j}{q_j(\ell_*)}.$$

The optimal investment strategy therefore corresponds to maximizing the quantity $< \ln(S_T/S_0) >$; note that one obtains a very different result by maximizing $< S_T/S_0 >$, that is, by seeking the maximum of $< X(\ell) >=< 1 + (u - 1)\ell >$.

To illustrate the great difference between the two strategies, let us consider the case with $M = 2$, $u^{(1)} = 2$ and $u^{(2)} = 0$ with probabilities respectively $p_1 = p$ and $p_2 = 1 - p$, in this type of market buying a stock is equivalent to playing a lottery in which, if you win, you get double the price of the ticket. Since $< 1 + (u - 1)\ell >= \ell(2p - 1)$, if $p > 1/2$ in the strategy where $< S_T/S_0 >$ is maximized, the optimal value of ℓ is 1. This is clearly a suicidal strategy: just one outcome of $u_t = u^{(2)} = 0$ (which will certainly happen eventually) is enough to wipe out the capital. With the logarithmic strategy, where $< \ln X(\ell) >$ is maximized, if $p > 1/2$ one obtains $\ell_* = 2p - 1$ (some is left in the bank) and

$$\langle \ln X(\ell_*) \rangle = \ln 2 + p \ln p + (1 - p) \ln(1 - p),$$

note that $h_p = -p \ln p - (1 - p) \ln(1 - p)$ is the Shannon entropy of the process $\{u_t\}$ (see Chap. 9) and $\max_p h_p = \ln 2$. As first noted by Kelly, the connection between entropy and growth rate is clear, in fact,

$$\langle \ln X(\ell_*) \rangle = \max_p h_p - h_p \ .$$

It is not difficult to repeat the calculations in the case of N different possible stocks: one must optimize the fractions $\ell_1, \ell_2, \dots, \ell_N$ where ℓ_k is the fraction of capital invested in the k-th stock. Equation (3.34) generalizes easily:

$$S_{t+1} = \Big[(1 - \ell_0) + \sum_{k=1}^{N} \ell_k u_{k,t}\Big] S_t,$$

where $1 - \ell_0 = 1 - (\ell_1 + \ell_2 + \ldots + \ell_N)$ is the fraction of capital left in the bank and $u_{k,t} = p_{k,t+1}/p_{k,t}$ is the ratio between the price of the k-th stock at time $t+1$ and at time t. One must therefore determine the $\ell_1^*, \ell_2^*, \ldots, \ell_N^*$ that maximize the quantity

$$\langle \ln\Big[(1-\ell_0) + \sum_{k=1}^{N} \ell_k u_k\Big] \rangle$$

Also in this case, one obtains an expression for $< \ln[(1-\ell_0^*) + \sum_{k=1}^{N} \ell_k^* u_k] >$ in terms of the Shannon entropy associated with the stock prices.

Not all finance experts agree with the previous strategy. According to the so-called utility theory, there is no "objective" strategy, that is, one independent of the beliefs and needs of the individual: each investor can decide (rationally) to maximize the expected value of their "utility function." In the case considered (bank plus stock market) one must maximize the expected value of the utility function $< F(X(\ell)) >$ where $F(X)$ is convex (convexity, i.e., $F'' < 0$, reflects the rationality of the investor; in short, all else being equal, it is preferable to earn more rather than less). The strategy based on the law of large numbers consists in assuming as utility function $F(X) = \ln X$; since $\ln X$ is a convex function, within the framework of utility theory this is a legitimate choice, but without any special status. For a detailed discussion, we refer the interested reader to the specialized literature. However, we note that the choice $F(X) = \ln X$ maximizes the gain and is an objective strategy, at least in the long run.

3.5.2 *Many Independent Causes Do Not Always Lead to the Gaussian: The Lognormal Distribution*

Let us consider the following multiplicative process:

$$m_n = x_n m_{n-1} = \Big[\prod_{j=1}^{n} x_j\Big] m_0 \tag{3.35}$$

where x_j are i.i.d. random variables, positive and bounded. Using the result from Sect. 3.3 we have that, to first approximation, the quantity $y_N = m_N/m_0$ for $N \gg 1$ has a lognormal probability distribution:

$$p_{Y_N}(y_N) \simeq p_{LN}(y_N) = \frac{1}{y_N\sqrt{2\pi C^2 N}} e^{-(\ln y_N - <\alpha> N)^2/(2C^2 N)}$$

where $< \alpha > = < \ln x >$ and C^2 is the variance of $\alpha = \ln x$. Obviously, the considerations previously made for large deviations still hold, so for extreme values of $\ln y_N/N - <\alpha>$ a more detailed treatment is needed in terms of the Cramer function, which depends on the probability density of x.

It is interesting that the lognormal distribution is present in many situations: from geology to biology and finance. For example, it describes reasonably accurately:

(a) the price of insurance against fires and industrial accidents;
(b) the number of sick days taken by workers in a company over a given period;
(c) the number of bacteria in a colony;
(d) the size of soil particles;
(e) the mass of pieces of coal (and other minerals) extracted in mines;
(f) the density of energy dissipated in developed turbulence.

There is no universally accepted explanation for this widespread presence of the lognormal distribution. However, it is possible to provide a plausibility argument based on multiplicative processes (3.35), which are quite common. As an example, we can think of m_n as the mass of a rock in the mountains. We can assume that the rocks present on the surface are the result of fragmentations that occur annually: water seeps into cracks and during the winter, as it freezes, the rock may break, for example, it remains intact with probability p or splits into two equal pieces with probability $1-p$, so we have $x = 1$ with probability p and $x = 1/2$ with probability $1 - p$. Thinking of this process repeated over many years, (3.35) is a model for the formation of rocks in the mountains or grains of sand, whose mass follows, to a good approximation, the lognormal distribution.

We can formalize the plausibility of the multiplicative process (3.35) with the following probabilistic model. Let $N_k(x)$ denote the number of particles (rocks) of mass less than x after k fragmentation events. Let $M_k(x) = E(N_k(x))$ and $B_k(x|y)$ be the average number of particles of mass less than x generated in the k-th fragmentation from particles of mass y, we have

$$M_k(y) = \int_0^\infty B_k(y|x) dM_{k-1}(x). \tag{3.36}$$

Assuming that the fragmentation process is independent of scale[13] that is, $B_k(x|y)$ depends only on the ratio y/x:

$$B_k(x|y) = C_k\left(\frac{y}{x}\right),$$

we have

$$M_k(y) = \int_0^\infty C_k\left(\frac{y}{x}\right) dM_{k-1}(x).$$

[13] This assumption is not always realistic: in many cases, smaller particles break more rarely. The validity of this hypothesis is generally restricted to values of the ratio y/x within a suitable interval.

Differentiating with respect to y the previous equation and noting that $dM_k(x) = const.p_k(x)dx$ where $p_k(x)$ is the probability density of x after k fragmentations, we have

$$p_k(y) = \int_0^\infty g_k\left(\frac{y}{x}\right) p_{k-1}(x) \frac{1}{x} dx, \tag{3.37}$$

where g_k is the derivative of C_k.

It is not difficult to show that the previous fragmentation process is nothing but a multiplicative process. Let us consider two independent variables x_1 and x_2 with probability distributions p_1 and p_2, respectively. The probability distribution for the variable $z = x_1 x_2$ is given by

$$p_Z(z) = \int\int p_1(x_1) p_2(x_2) \delta(z - x_1 x_2) dx_1 dx_2,$$

using the well-known properties of the Dirac delta, we have

$$p_Z(z) = \int p_1(x_1) p_2\left(\frac{z}{x_1}\right) \frac{1}{x_1} dx_1.$$

Equation (3.37) is nothing but the previous formula for the probability distribution of the product of two independent variables, and therefore the multiplicative process (3.31) is justified under the hypothesis of scale invariance, that is, $B_k(x|y) = C_k(y/x)$.

Exercises

3.1 Consider the Buffon's needle experiment discussed in Chap. 2. Give an estimate of the number of trials necessary to have a probability less than 10^{-3} that the percentage error exceeds 10^{-3}.

For the numerical solution, it is necessary to consult a table for the integral

$$\frac{1}{\sqrt{2\pi}} \int_0^x e^{-z^2/2} dz,$$

these tables can be found in many probability books, for example, in Gnedenko's book.

3.2 Let x_1, x_2, and x_3 be independent and uniformly distributed in $[-1, 1]$; show that the probability density of $z = x_1 + x_2$ and $q = x_1 + x_2 + x_3$ are respectively

$$p_Z(z) = \begin{cases} (2 - |z|)/4 \ \ if \ |z| < 2 \\ 0 \ \ if \ |z| > 2 \end{cases}$$

$$p_Q(q) = \begin{cases} 0 \ \ if \ |q| > 3 \\ (3 - |q|)^2/16 \ \ if \ 1 \le |q| \le 3 \\ (3 - q^2)/8 \ \ if \ 0 \le |q| \le 1 \end{cases}$$

3.3 Let $a_1, \ldots, a_N$ be i.i.d. variables that take the value 0 or 1 with equal probability 1/2. Show that in the limit $N \to \infty$ the variable

$$x_N = \sum_{j=1}^{N} a_j 2^{-j}$$

is uniformly distributed in [0, 1].

Repeat the problem in the case where the a_j take values $0, 1, \ldots, M-1$ with equal probability $1/M$ and

$$x_N = \sum_{j=1}^{N} a_j M^{-j}.$$

3.4 Given $x_1, \ldots, x_N$ independent Gaussian variables such that x_k has mean m_k and variance σ_k^2, using characteristic functions show that the variable

$$y = \sum_{j=1}^{N} x_j$$

is Gaussian with mean $\sum_k m_k$ and variance $\sum_k \sigma_k^2$.

3.5 Given $x_1, \ldots, x_N$ non-independent Gaussian variables such that $< x_k >= 0$, $\sigma_k^2 = 1$ and $< x_k x_{k+n} >= g(n)$, show that the variable

$$y = \sum_{j=1}^{N} x_j$$

is Gaussian with zero mean and variance

$$N + 2 \sum_{n=1}^{N-1} (N-n) g(n).$$

3.6 Within the canonical ensemble at temperature T, consider a gas of N particles of mass m moving on a square of side L. Calculate the probability density $p(E)$ of the energy E, find the value E^* at which the maximum occurs, and show that

$$\lim_{N \to \infty} \frac{E^*}{< E >} = 1,$$

and furthermore, for every $n > 0$,

$$\lim_{N\to\infty} \frac{< E^n >}{< E >^n} = 1.$$

3.7 Within the canonical ensemble at temperature T, consider a one-dimensional system of N oscillators with Hamiltonian

$$H = \sum_{n=1}^{N} \frac{p_n^2}{2m} + \sum_{n=1}^{N}\sum_{j=1}^{N} A_{n,j} x_j x_n$$

where the matrix $\{A_{n,j}\}$ is symmetric and its eigenvalues are positive. Calculate the probability density of the energy E, and compare the result with that of the previous exercise.

3.8 Within the canonical ensemble at temperature T, consider a one-dimensional system consisting of N particles of mass m (referred to as light) and one (heavy) particle of mass M. The light particles can pass through each other and move between $x = 0$ and the position X of the heavy particle, which is subject to a potential FX. The Hamiltonian of the system is

$$H = \sum_{n=1}^{N} \frac{p_n^2}{2m} + \frac{P_0^2}{2M} + FX,$$

where P_0 is the momentum of the heavy particle.

(a) Calculate the probability density of the position of the heavy particle;
(b) follow the system over time and sample X every time one of the light particles touches the boundary $x = 0$, calculate the new probability density.

Note: this system is a model of a thermometer. To perform a dynamic simulation, an algorithm is needed that takes into account the "thermal bath" at $x = 0$, see *ES9.2*. The problem of determining the temperature from empirical data and its uncertainty is discussed in M. Falcioni, D. Villamaina, A. Vulpiani, A. Puglisi and A. Sarracino, "Estimate of temperature and its uncertainty in small systems", Am. J. Phys. **79**, 777 (2011).

3.9 Let $x_1, \ldots, x_N$ be i.i.d. Cauchy variables with density

$$p_X(x) = \frac{1}{\pi(1 + x^2)}.$$

Calculate the probability density of the variable

$$y = \frac{1}{N} \sum_{j=1}^{N} x_j .$$

Discuss why in the limit $N \gg 1$ the law of large numbers does not apply.

3.10 Given an algorithm that generates independent random numbers uniformly distributed in [0, 1], discuss how to use the idea of Buffon's needle to numerically calculate an integral of the type

$$\int_0^1 f(x) dx$$

where $f(x)$ in the interval [0, 1] is between 0 and 1.

3.11 Recalling that the distribution of a sum of N Poisson variables is still Poisson, using the CLT show that

$$\lim_{N\to\infty} e^{-N}\left(1 + N + \frac{N^2}{2!} + \frac{N^3}{3!} + \ldots + \frac{N^N}{N!}\right) = \frac{1}{2},$$

and furthermore

$$\lim_{N\to\infty} e^{-N} \sum_{k=[N+x_1\sqrt{N}]}^{[N+x_2\sqrt{N}]} \frac{N^k}{k!} = \frac{1}{\sqrt{2\pi}} \int_{x_1}^{x_2} e^{-x^2/2} dx,$$

where [ ] denotes the integer part.

Recommended Readings

A clear treatment of the limit theorems can be found in the book by Gnedenko (see Chap. 2) or in

A. Renyi, *Probability Theory* (Dover Publications, New York, 2007) in French *Calcul des probabilités* (Jacques Gabay Ed. 2000)

For limit theorems for non-independent variables:

I.A. Ibragimov, Yu.V. Linnik, *Independent and Stationary Sequences of Random Variables*. Wolters-Noordhoff Series of Monographs and Textbooks on Pure and Applied Mathematics (Walterss-Noordhoff, Groningen, 1971)

For an introductory discussion on large deviations:

H. Touchette, The large deviation approach to statistical mechanics. Phys. Rep. **478**, 1 (2009)

For an introduction to fluctuation theory in statistical mechanics, see the books by Reichl and Peliti cited in Chap. 2.

A classic book on the importance of limit theorems for statistical mechanics:

A.I. Khinchin, *Mathematical Foundations of Statistical Mechanics* (Dover Publications, New York, 1960)

For investment strategies in finance and limit theorems:

T.M. Cover, J.A. Thomas, *Elements of Information Theory* (Wiley, New York, 1991)

For a discussion on the lognormal distribution:

E.A. Bender, *An Introduction to Mathematical Modeling* (Wiley, New York, 1978)

Chapter 4
Brownian Motion: First Encounter with Stochastic Processes

4.1 Observations

The study of Brownian motion began with the observations of the Scottish botanist Robert Brown in 1827. Brown made fundamental contributions to biology, including the discovery of the cell nucleus, but paradoxically his name became famous for recognizing the purely physical origin of the phenomenon that bears his name.

During the summer of 1827, while studying the role of pollen in the fertilization process of plants, Brown observed that pollen grains (a few microns in diameter) seen under the microscope exhibited an irregular and incessant motion. Certainly, Brown was not the first to observe this phenomenon, but he was the first to seriously question its origin.

In 1828, Brown published a booklet entitled "*A brief account of microscopical observations made in the months of June, July and August, 1827, on the particles contained in the pollen of plants; and on the general existence of active molecules in organic and inorganic bodies*" about his observations from the previous summer, in which he showed that the origin of the motion "is not due to currents in the fluid or its evaporation, but is inherent to the particles themselves." Brown then considered other particles, both pollen and others of organic origin, and then also of inorganic origin (for example, dust obtained from glass), and for all he found the same type of behavior. Brown's great merit was not so much in having observed Brownian motion, but in having taken it out of the purely biological context and correctly framing it as a physical phenomenon.

Brown's work aroused interest and a series of possible hypotheses about the origin of the observed motion, partly generated by the ambiguous use of the term "active molecules" in his booklet. To respond to these hypotheses and to clarify the non-biological origin of the phenomenon, Brown wrote a second memoir the following year. After the initial interest, Brown's work was long neglected, although it did attract the interest of M. Faraday, who repeated the experiment and confirmed

G. Boffetta, A. Vulpiani, *Probability in Physics*, UNITEXT for Physics,
https://doi.org/10.1007/978-3-032-10407-6_4

Brown's observations. In the second half of the nineteenth century, it began to be correctly thought that the origin of Brownian motion lay in collisions with the molecules of the fluid. In particular, Giovanni Cantoni suggested in the conclusions of one of his articles that "*Brownian motion provides us with one of the most beautiful and direct experimental demonstrations of the fundamental principles of the mechanical theory of heat, manifesting that constant vibratory state that must exist in both liquids and solids even when their temperature does not change*."

Towards the end of the century, the French physicist Louis-Georges Gouy conducted a series of very precise and systematic experiments on Brownian motion, varying the temperature and viscosity of the fluid and the size of the grains. From these experiments, he found, among other things, that the motion is more active at higher temperatures, lower viscosities, and for smaller particles, while the chemical nature of the particles does not matter. Gouy also realized that Brownian motion is in apparent contradiction with the second law of thermodynamics, according to which it is not possible to extract work from a single heat source. The motion of the pollen grain shows, in fact, that the molecules of the fluid can (apparently) produce work, even if only erratically. These difficulties were also recognized by Poincaré, who in 1904 commented that Brownian motion "*is the opposite of Carnot's principle: to see the world run backwards, the extremely fine vision of Maxwell's demon is not necessary, but a microscope is sufficient*."

4.2 The Theory: Einstein and Smoluchowski

The fundamental turning point in the study of Brownian motion occurred at the beginning of the twentieth century with the theoretical contributions of Einstein and Smoluchowski and the subsequent reformulation by Langevin. These works mark the birth of the theory of stochastic processes.

In his annus mirabilis 1905, the 26-year-old Einstein, still an employee of the famous patent office in Bern, published 5 papers, all of fundamental importance for the development of physics. Two articles mark the birth of special relativity; the article on the photoelectric effect, for which Einstein would receive the Nobel Prize in 1921, laid the foundations of quantum mechanics; one article concerned the size of molecules, and finally, the first article on Brownian motion.

Compared to the pillars of modern physics of relativity and quantum mechanics, the articles on molecular dimensions and on Brownian motion may seem minor. In reality, Einstein's interest in this field was just as noble and fundamental, as it was motivated by the desire to demonstrate the existence of atoms. In fact, it should be remembered that at the beginning of the twentieth century, the reality of atoms was not yet fully accepted and still had many opponents. Among the most famous were Mach and the Nobel Prize-winning chemist Ostwald (who, paradoxically, introduced the term "mole," establishing himself among the founders of modern physical chemistry). As he writes in his scientific autobiography, Einstein wanted to "*find facts that could guarantee as much as possible the existence of atoms*".

The purpose of his work is clearly stated at the beginning of the 1905 article: "*In this article it will be shown that, in accordance with the molecular kinetic theory of heat, bodies of dimensions visible under the microscope suspended in a liquid are endowed with movements of such magnitude as to be easily observed*". In other words, for Einstein, Brownian motion becomes a "natural microscope" to directly observe the atomic world.

The fundamental result of Einstein's 1905 work is the expression, known as the Einstein-Smoluchowski relation, for the diffusion coefficient of the pollen grain in terms of quantities from the microscopic world, in particular Avogadro's number N_A. We thus have, as mentioned, a mathematical relation that links the macroscopic world (the pollen grain) to the unobservable microscopic world (the molecules). In the following years, Einstein published a couple more papers on Brownian motion in which he explicitly drew the reader's attention to the correct quantities to measure. In particular, Einstein observed that the average velocity of the grain over a time interval τ is not a good observable, since its value is inversely proportional to $\sqrt{\tau}$, and this explains the experimental difficulties in obtaining an estimate of the grain's velocity to compare with theory. The "correct" observable to measure, Einstein suggests, is the mean square displacement of the particle, which turns out to be proportional to time via the diffusion coefficient.

Smoluchowski published his first work on Brownian motion in 1906, citing Einstein's work, which is "*in perfect agreement with what I obtained a few years ago with a completely different, simpler, more direct, and perhaps more convincing reasoning than Einstein's*". Smoluchowski's reasoning is based on an approach to the problem in terms of kinetic theory (appropriately simplified), in which the microscopic process of the grain's collision with the molecules is stripped of its physical details and replaced with a simple random process. That is, the collision is considered a random event, similar to tossing a coin. In this way, Smoluchowski obtains an expression for the diffusion coefficient identical to Einstein's (apart from a small inaccuracy in the numerical factor).

We conclude this brief historical introduction by recalling that the Einstein-Smoluchowski law was also derived by the Australian physicist William Sutherland, who published a paper in 1905, a few months before Einstein. The reason why Sutherland's contribution is still little known is not entirely clear, but probably among the causes, a non-negligible role is played by the fact that at the beginning of the twentieth century, theoretical physics was essentially German, while the Anglo-Saxon world, to which Sutherland belonged, was more advanced in experimental physics. Certainly, it should also be considered that, among other possible factors, it must not have been easy for a "regular" good physicist to compete for a simultaneous discovery with a giant of science such as Einstein.

4.3 Langevin's Derivation and the Nobel Prize to Perrin

In 1908, Paul Langevin proposed an independent derivation of the result of Einstein and Smoluchowski with a method that provides the first example of a *stochastic differential equation*. Langevin's derivation is simple and illuminating, so we report it here below.

Langevin's reasoning starts from a dynamical model in which the forces acting on the pollen grain are of two types: a macroscopic and systematic (deterministic) force due to friction with the fluid, and a microscopic stochastic force due to collisions with the molecules. Newton's law for the grain is thus written (for simplicity, we write here only the component in one direction)

$$m\frac{dv}{dt} = -6\pi a\mu v + \xi \tag{4.1}$$

where m and a are the mass and radius of the grain (assumed spherical), $v = dx/dt$ its instantaneous velocity, and μ is the viscosity of the fluid. The first term on the right in (4.1) is the (so-called Stokes) friction force on a spherical body moving in a fluid, while ξ represents the force due to collisions with the molecules.

If we neglect ξ, we can integrate (4.1) and obtain that the velocity tends to zero, due to friction, with a characteristic relaxation time given by $\tau = m/(6\pi\mu a)$. For a micron-sized grain placed in water at room temperature, this Stokes time is very short, $\tau = O(10^{-7})$ seconds, but it is still very large compared to the typical times of collisions with molecules, which are $O(10^{-11})$ seconds. Therefore, we can assume that on the characteristic timescales of the grain, the force ξ is an uncorrelated noise in time and independent of the position of the grain. Multiplying (4.1) by x and averaging over many collisions one obtains

$$\frac{1}{2}\frac{d^2}{dt^2}\langle x^2\rangle - \langle v^2\rangle = -\frac{1}{2\tau}\frac{d}{dt}\langle x^2\rangle + \frac{1}{m}\langle x\xi\rangle \tag{4.2}$$

The last term represents the correlation between the position of the grain x and the force due to the impact of the molecules. For the reasons explained earlier, we can reasonably assume that the collisions of the molecules do not depend on the position of the grain and therefore, since we have collisions from all directions, on average we have $\langle x\xi\rangle = 0$.

At this point, the crucial step in the derivation is to assume that the grain is in thermodynamic equilibrium with the molecules. Note that this is a very strong assumption, given the large difference in size between the grain and the molecules, but it is precisely thanks to this bold hypothesis by Einstein that the problem can be solved. Formally, this hypothesis implies that we can apply the equipartition principle of energy to the grain and write $\langle v^2\rangle = k_BT/m$. In this way, Eq. (4.2) becomes an elementary differential equation for the variable $\langle x^2\rangle$ which, when integrated (assuming that the initial position is $x(0) = 0$), gives:

$$\langle x^2(t) \rangle = \frac{2k_B T}{m} \tau^2 \left[\frac{t}{\tau} - (1 - e^{-t/\tau}) \right] \quad (4.3)$$

In the limit of times much longer than the relaxation time, $t \gg \tau$, in the solution (4.3) only the first term survives and one obtains

$$\langle x^2(t) \rangle = \frac{2k_B T}{m} \tau t = 2Dt \quad (4.4)$$

This is the *diffusion law* for Brownian motion: on average, the particle does not move, $\langle x \rangle = 0$, because it receives as many collisions from the right as from the left. On the other hand, the mean *quadratic* displacement (which, being a square, is not affected by the sign) is not zero and grows linearly with time.[1]

The proportionality constant D in (4.4) is called the *diffusion coefficient* and is expressed by the Einstein-Smoluchowski relation as

$$D = \frac{k_B T}{6\pi a \mu} \quad (4.5)$$

The importance of the relation (4.5) is that it connects the macroscopic quantity D, which can be determined from experimental observables as explained above, with microscopic quantities such as the Boltzmann constant k_B and Avogadro's number $N_A = R/k_B$ (R is the gas constant). In other words, (4.5) unambiguously links the microscopic world (the molecules) with the macroscopic world (the grain) and allows one to determine a quantity of the former (Avogadro's number) by measuring the latter (the diffusion coefficient). The crucial point of the derivation is that the pollen grain is required to obey, simultaneously, both macroscopic hydrodynamics (Stokes' law) and kinetic theory (equipartition), thus providing a bridge to the world of molecules. From this apparent contradiction, the bold and brilliant idea of Einstein and Smoluchowski, comes the definitive certainty of the correctness of the atomic hypothesis (Fig. 4.1).

In 1908, the French physicist Jean Baptiste Perrin carried out a series of quantitative experiments on Brownian motion in order to verify the accuracy of the molecular hypotheses of Einstein's derivation and to measure the value of Avogadro's number. For his experiments, Perrin used emulsions of gamboge treated with alcohol and centrifuged to obtain granules of known diameter. These were observed using an immersion microscope, which increased the resolution and

[1] A simple way to understand the phenomenology of Brownian motion is to pour a drop of dye into a glass of water. Taking care not to stir the water and waiting long enough, we will observe that the position of the center of the spot does not change (that is, $\langle x \rangle = 0$) but the area of the spot increases over time. If we measured it, we would find that the area grows proportionally to t, as predicted by (4.4).

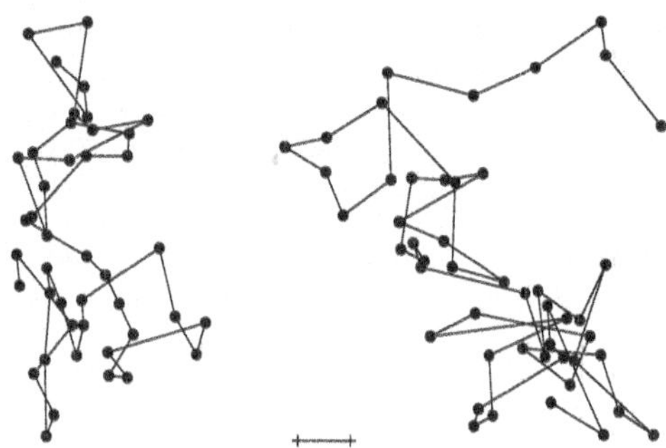

Fig. 4.1 Two examples of tracks of colloidal particles of radius 0.52 μm as observed under the microscope by Perrin. The segment represents a distance of 6.4 μm and the positions are shown every 30 seconds

Table 4.1 List of Perrin's determinations of Avogadro's number N_A from experiments on Brownian motion with different classes of grains

Nature of the emulsion	Radius of Grains (μm)	Mass of Grains ($g/10^{15}$)	$N_A/10^{22}$
Gamboge	0.212	48	69.5
Gamboge	0.367	246	68.8
Gamboge	0.50	600	80
Mastic	0.52	650	72.5
Mastic	5.50	750,000	78

allowed better temperature control. Using grains of very different diameters, as indicated in Table 4.1 and also of other materials, Perrin obtained a series of values for Avogadro's number that were close to each other and compatible with the value determined by other methods. Perrin concluded: "*this remarkable agreement proves the rigorous accuracy of Einstein's formula and spectacularly confirms the molecular theory*".

The experimental verification of Einstein's theory convinced even the last skeptics about the reality of atoms. For his research on the molecular nature of matter, Perrin received the Nobel Prize in Physics in 1926. The story of his research on Brownian motion is recounted in his book *Les Atomes*, whose reading is a valuable source of ideas and knowledge for every physics student.

4.4 A Simple Stochastic Model for Brownian Motion

Langevin's derivation has the merit of simplicity and explains the physical origin of the diffusive law (4.4). This can be obtained formally using the theory of continuous-time stochastic processes. It is possible to get an idea of the approach with the theory of stochastic processes by means of a simple discrete-time model.

Let us consider a one-dimensional model in which the pollen grain can move along the x axis. At each discrete time $t = n\Delta t$, the position and velocity of the grain will be given by the quantities x_n and v_n, which evolve according to the stochastic rules

$$x_{n+1} = x_n + v_n \Delta t \tag{4.6}$$

$$v_{n+1} = a v_n + b w_n \tag{4.7}$$

where a and b are constants that will be determined consistently. Equation (4.7) is the discrete version of (4.1), in which the first term represents the deterministic evolution and the second the stochastic force due to the molecules. The random variables w_n are assumed to be independent at each step and distributed according to a normalized Gaussian, $N(0, 1)$ using the notation of Chap. 3.

Let us assume that the initial velocity is normally distributed, that is, its probability density is Gaussian with mean $\langle v_0 \rangle$ and variance σ_0^2, and we denote this distribution by $N(\langle v_0 \rangle, \sigma_0)$. After one step, v_1 will still be normally distributed (since a linear combination of Gaussian variables is itself Gaussian) with $N(\langle v_1 \rangle, \sigma_1)$. Using (4.7) we have

$$\langle v_1 \rangle = a \langle v_0 \rangle$$

while

$$\langle v_1^2 \rangle = a^2 \langle v_0^2 \rangle + b^2 \langle w_0^2 \rangle + 2ab \langle v_0 w_0 \rangle = a^2 \langle v_0^2 \rangle + b^2$$

Repeating the calculation for a generic n we obtain

$$\langle v_n \rangle = a^n \langle v_0 \rangle \tag{4.8}$$

while for $\sigma_n^2 = \langle v_n^2 \rangle - \langle v_n \rangle^2$ we have

$$\sigma_{n+1}^2 = a^2 \sigma_n^2 + b^2 \tag{4.9}$$

Now assume that $0 < a < 1$; therefore, for $n \to \infty$ we have $\langle v_n \rangle \to 0$, while from (4.9) we obtain

$$\lim_{n \to \infty} \sigma_n^2 = \frac{b^2}{1 - a^2} \equiv \langle v^2 \rangle. \tag{4.10}$$

Therefore, the velocity v_n of the grain at time n is Gaussian distributed and his probability distribution quickly tends to the asymptotic distribution $N(0, \langle v^2 \rangle)$. In fact, we can rewrite (4.9) for the distance from the asymptotic value $\delta_n = \sigma_n^2 - b^2/(1 - a^2)$ as $\delta_{n+1} = a^2 \delta_n$.

In physical terms, the value of a defines a characteristic time beyond which the velocity distribution becomes practically stationary. Therefore, since we are interested in the motion of the grain over "long" times, it is reasonable to assume the velocity distribution given by the asymptotic one $N(0, \langle v^2 \rangle)$ and to study only the random walk, described by (4.6). After n steps, the displacement of the grain will be given by

$$\Delta_n = x_n - x_0 = \Delta t \sum_{j=0}^{n-1} v_j$$

with obviously $\langle \Delta_n \rangle = 0$, while for the quadratic displacement

$$\langle \Delta_n^2 \rangle = (\Delta t)^2 \sum_{j=0}^{n-1} \langle v_j^2 \rangle + 2(\Delta t)^2 \sum_{j=0}^{n-1} \sum_{k=1}^{n-j-1} \langle v_j v_{j+k} \rangle$$

For the second term, we observe that $\langle v_j v_{j+k} \rangle = a^k \langle v^2 \rangle$ (in fact, by multiplying (4.7) by v_n and taking the average, one obtains $\langle v_n v_{n+1} \rangle = a \langle v^2 \rangle$; instead, multiplying by v_{n-1}, taking the average, and using the previous result, we obtain $\langle v_{n-1} v_{n+1} \rangle = a^2 \langle v^2 \rangle$, and so on. Recalling that $\sum_{k=1}^{N} a^k = a(1-a^N)/(1-a)$, we obtain

$$\langle \Delta_n^2 \rangle = (\Delta t)^2 n \langle v^2 \rangle + 2(\Delta t)^2 \langle v^2 \rangle \frac{a}{1-a} \left[n - \frac{1-a^n}{1-a} \right].$$

Now let us fix the values of the coefficients a and b based on consistency considerations. For $b = 0$, Eq. (4.7) must be the discrete version of (4.1) with $f = 0$, that is, $dv/dt = -v/\tau$. To first order in Δt we must have $v(t+\Delta t) - v(t) = -v(t)\Delta t/\tau = v_{n+1} - v_n = a v_n$ and therefore

$$a = 1 - \frac{\Delta t}{\tau}$$

while from (4.10) we have

$$b = \sqrt{\frac{2\Delta t \langle v^2 \rangle}{\tau}}.$$

We can now calculate the diffusion coefficient defined by (4.4). In our case we obtain

$$D = \lim_{n \to \infty} \frac{\langle \Delta_n^2 \rangle}{2n\Delta t} = \tau \langle v^2 \rangle. \tag{4.11}$$

The characteristic time τ is given by the Stokes formula $\tau = m/(6\pi\mu R)$ (R is the radius of the particle and μ the viscosity of the fluid), while the value of $\langle v^2 \rangle$ is determined by the equipartition of energy $\langle v^2 \rangle = k_B T/m$, as described in the previous chapter. Substituting into (4.11) we recover the Einstein-Smoluchowski formula

$$D = \frac{k_B T}{6\pi \mu R}. \tag{4.12}$$

This derivation, although very simplified, still preserves the important properties of diffusive motion and allows us to understand some fundamental aspects. In particular, we learn that since the distribution of velocities v_n rapidly converges to the asymptotic distribution, we can obtain the same result (4.12) by considering directly only the equation for the random walk (4.6) with a given (Gaussian) distribution of velocities v_n.

Some details of the model used (for example, the Gaussianity of w_n) are inessential. The reader can check with simple calculations (see the Exercises) that by assuming (4.6) where v_n has a probability distribution $p_n(v)$ and correlations that decay rapidly, then

$$\frac{\langle (x_n - x_0)^2 \rangle}{2n} \to_{n\to\infty} \frac{\langle v^2 \rangle}{2} \Delta t^2 + \sum_{j=1}^{\infty} \langle v_j v_0 \rangle \Delta t^2.$$

Moreover, assuming (4.7) with $\{w_n\}$ independent variables distributed according to a probability distribution $g(w)$, (4.8) and (4.9) are still valid and for $n \to \infty$, $p_n(v) \to p(v)$, not necessarily Gaussian, solution of the equation

$$p(v) = \int p(v')g(w)\delta(v - av' - bw)dv'dw.$$

4.5 An Even Simpler Model: The Random Walk

As mentioned earlier, in the model (4.6)–(4.7) the velocities quickly reach the asymptotic distribution and the problem *decouples*, in the sense that one can consider the evolution of the position (4.6) in a given velocity distribution (Fig. 4.2).

To better illustrate this concept, let us consider a further simplified version in which the velocity can take only 2 discrete values: $v_n = +v$ and $v_n = -v$. Clearly, the position of the grain x_n evolving according to (4.6) will move on a discrete set of points separated by $\Delta x = v\Delta t$. If we take the starting point of the grain as the origin (i.e., $X(0) = 0$), its position after a time $t = n\Delta t$ will be

$$X(t) = \Delta t \sum_{i=1}^{n} V_i$$

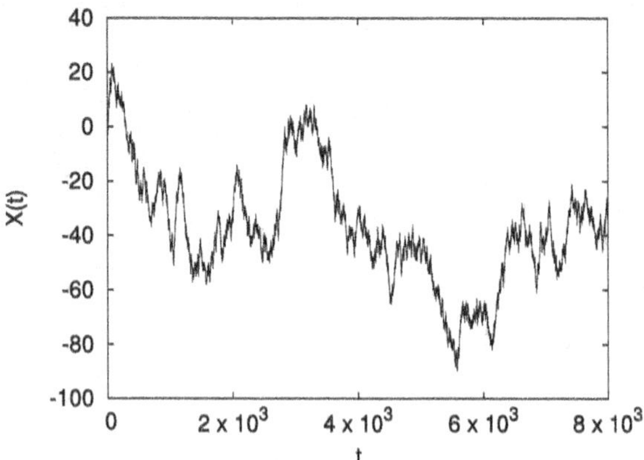

Fig. 4.2 A realization of $n = 8000$ steps of a random walk with $\Delta x = \Delta t = 1$ and $X(0) = 0$

The probability $p_t(X = x)$ of finding the grain at position $x = k\Delta X$ at time $t = n\Delta t$ will be given by the probability $p_n(k)$ of making $(n+k)/2$ jumps with $V_i = +v$ and $(n-k)/2$ jumps with $V_i = -v$ out of a total of n jumps. Obviously, the order in which the positive and negative jumps are made does not matter, and therefore we have

$$p_t(x) = \frac{1}{2^n}\binom{n}{\frac{n+k}{2}}.$$

For large values of n and k, that is, after a large number of steps, we can use Stirling's approximation $\log n! \simeq n\log(n) - n$ and obtain, returning to the physical variables:

$$p_t(x) = \frac{1}{\sqrt{4\pi Dt}} e^{-\frac{x^2}{4Dt}} \tag{4.13}$$

therefore, the position of the grain will have a Gaussian distribution with zero mean and variance given by $\langle x^2(t)\rangle = 2Dt$ with

$$D = \frac{1}{2}v^2\Delta t.$$

The diffusion coefficient D is still in the form (4.11), that is, the product of the velocity variance by a decorrelation time (which in this case is $\Delta t/2$).

Let us also note that the results of this chapter confirm what was discussed in Chap. 3 regarding the Central Limit Theorem, for which the Gaussian distribution (4.13) appears in a very general way.

Exercises

4.1 Let us consider a variant of the discrete-time Brownian motion model discussed in Sect. 4.4

$$v_{n+1} = av_n + z_n, \quad x_{n+1} = x_n + v_n$$

where $|a| < 1$ and the $\{z_n\}$ are i.i.d. with zero mean and known, non-Gaussian probability density $p_Z(z)$:

(a) assuming that for $n \to \infty$ the probability density of v_n does not change, show that v_n is not Gaussian;
(b) show that in the limit $n \gg 1$ the probability density of x_n is Gaussian and $< x_n^2 > \simeq 2Dn$, calculate D.

4.2 Consider the discrete-time stochastic process

$$x_{n+1} = ax_n + z_n$$

where $|a| < 1$ and the $\{z_j\}$ are i.i.d. with known probability density $p_Z(z)$, which for $|z| \to \infty$ decays rapidly to zero so that all moments exist.

Show that in the limit $n \to \infty$ the probability density of x_n tends to a limiting function, and discuss the rate of convergence.

Hint: use characteristic functions and cumulants.

Suggested Readings

The works of A. Einstein on Brownian motion have been translated into English and collected in the volume *Investigation on the Theory of the Brownian Motion* (Dover Publications, New York, 1956)

A classic on Brownian motion (and more)

J. Perrin, *Les Atomes* (Alcan, Paris, 1913)

The original article by Langevin:

P. Langevin, Sur la theorie du mouvement brownien. C. R. Acad. Sci. (Paris) **146**, 530 (1908) translated into English in Am. J. Phys. **65**, 1079 (1997)

For a review, partly historical, on Brownian motion with a detailed analysis of the works of Einstein, Smoluchowski, and Perrin, see

S. Chandrasekhar, Stochastic problems in physics and astronomy. Rev. Mod. Phys. **15**, 1 (1943)

Chapter 5
Discrete Stochastic Processes: Markov Chains

Historically, one of the first studies related to **stochastic processes** is due to the mathematician Bachelier, who in 1900 used the random walk model for the analysis of financial markets. The idea was then developed a few years later, with the works of Einstein, Smoluchowski, and Langevin and their applications to the study of Brownian motion, as discussed in the previous chapter.

A stochastic process is defined by a set of random variables $X_t \in \Omega$ that are functions of a parameter t, which generally represents time. In other words, a stochastic process is given by the time evolution of a random variable. In the case of a deterministic system, the evolution is given by a rule (differential equation or discrete map) that uniquely determines the value of the variable at a later time, given the state at the current time t. Conversely, in the case of stochastic processes, the evolution of the probability is specified. That is, given the current state, the system at a later time may be found in a set of different states, according to probabilistic rules.[1] Clearly, these rules in general may depend both on the state in which the system is found and on its past history. In the case where they depend only on the state of the system at time t (and not on the previous history), we speak of **Markov processes**.[2] We can say that a Markovian process is, in the probabilistic context, in some respects similar to a deterministic system since the future of the system is determined (in a statistical sense) only by the present state. Many processes of physical interest are, to a good approximation, *Markovian*, and this is fortunate since the study of non-Markovian processes is very complicated. In this book, therefore, we will deal exclusively with Markovian processes.

The simplest case, which will be treated in this chapter, is that of **Markov chains** in which both the random variable and time take discrete values. Markov chains

---

[1] We note that, almost always, the probabilistic nature of the evolution in stochastic processes reflects our ignorance regarding some underlying causes which, although deterministic, are modeled by means of random variables.

[2] In honor of the Russian mathematician Andrei Andreievich Markov (1856–1922).

G. Boffetta, A. Vulpiani, *Probability in Physics*, UNITEXT for Physics,
https://doi.org/10.1007/978-3-032-10407-6_5

with a finite number of states were introduced by Markov in 1906 and were later generalized to the case of an infinite number of states by Kolmogorov in 1936.

In the case where the states are discrete but time is continuous, $t \in \mathbb{R}$, the evolution of the probability is given by the so-called **master equation** introduced by Pauli in 1928 (in the study of relaxation to equilibrium in quantum systems) and later used by Uhlenbeck for the study of fluctuations in cosmic rays. A fundamental characteristic of a process described by a master equation is that, given the discrete nature of the states, for a small time interval ($\delta t \to 0$), the probability of not changing state is $1 - O(\delta t)$, much greater than the probability of changing it, which is $O(\delta t)$. One of the natural applications of this approach is in the study of population dynamics, where the discrete nature of the random variable is clearly linked to the presence of individuals.

Finally, in the case where both the states and time are continuous, we are in the realm of the **Fokker-Planck equation**: in this case, the state changes for any interval δt, but the change is small if δt is small, in a sense that we will define more precisely. The Fokker-Planck equation was already introduced by Einstein and Smoluchowski in 1906 and subsequently by Fokker in 1914. In 1917, Planck derived the equation in a general form, while the theory was later formalized by Kolmogorov in 1931. The Fokker-Planck equation, being a partial differential equation, is generally more tractable from a mathematical point of view than the corresponding discrete equations (Markov chains or master equation). For this reason, in many cases the Fokker-Planck equation is used as an approximate, "coarse-grained" description, even when the process is intrinsically discrete.

All these approaches, discrete or continuous, have in common the fact that they describe the deterministic (discrete or continuous) time evolution of the probability of finding the system in a certain state at a certain time. An alternative description is based on the time evolution of a single realization of the stochastic variable X_t, described, in the continuous case where it is generally used, by a **stochastic differential equation**, that is, an equation that explicitly contains stochastic terms. The rest of this chapter is devoted to the study of Markov chains (and, to a lesser extent, the master equation), while the study of the Fokker-Planck equation will be addressed in the next chapter, together with some basic concepts related to stochastic differential equations.

5.1 Markov Chains

Let us consider the case of a random variable with discrete values, for example taking values in $\Omega = \mathbb{Z}$, as a function of a "time" variable also discrete, $t \in \mathbb{Z}$. As we have seen in the study of Brownian motion, the fundamental quantity to determine is the probability of finding the random variable X in the generic state i at time t

$$p_i(t) = P(X_t = i).$$

In general, we can think that the successive states of the random variable X_t are correlated and therefore the probability of being in state j at time $t+1$ depends on the previous s states, that is, it is given by the conditional probability $P(X_{t+1} = j|(X_t = i) \cap (X_{t-1} = i_1) \cap \ldots \cap (X_{t-s+1} = i_{s-1}))$. The degree of "memory" s determines the type of stochastic process.

In the simplest case where there are no correlations, the state X_{t+1} does not depend on the previous states and thus simply

$$
\begin{aligned}
&P(X_{t+1} = j|(X_t = i) \cap (X_{t-1} = i_1) \cap \ldots \cap (X_{t-s+1} = i_{s-1})) \\
&= P(X_{t+1} = i) = p_j(t+1).
\end{aligned}
$$

This is the case of independent processes (such as the case of repeated draws) in which $p_j(t)$ completely determines the process.

The next case, which is of particular importance, is that of Markov processes (or chains, since we are dealing with discrete processes) in which the state of the system depends only on the previous state:

$$
\begin{aligned}
&P(X_{t+1} = j|(X_t = i) \cap (X_{t-1} = i_1) \cap \ldots \cap (X_{t-s+1} = i_{s-1})) \\
&= P(X_{t+1} = j|X_t = i) \equiv W_{ij}(t),
\end{aligned}
$$

where $W_{ij}(t)$ represents the probability of transition from state i at time t to state j at the next time step and, for a Markov process, does not depend on the path that brought the variable X_t to i. If the process is **stationary**, that is, invariant with respect to the choice of the time origin, this probability becomes independent of t. In the following, unless otherwise indicated, we will consider the case of stationary processes.

A Markov process is therefore by definition memoryless (or with finite memory), in the sense that it depends only on its present state and not on the past ones: this property allows, as we will see later, to greatly simplify the treatment of problems. Let us also recall that Markov processes find many applications in fields ranging from statistical mechanics, to the study of chemical reactions and biological processes (for example in population theory), to the mathematical study of economic-financial processes, and even to computer science applications (for example, the web page ranking technique adopted by Google is based on a Markov chain) and to automatic systems for text and music generation.

Let us note that the important requirement for Markov processes is that the evolution has finite memory. In fact, if we suppose that the future state depends on the last r states X_k (the current X_t plus the previous $r-1$ states $X_{t-1}, X_{t-2}, \ldots, X_{t-r+1}$), it is sufficient to redefine a new chain with state Y_t constructed from the r ordered states of X, $Y_t = (X_t, X_{t-1}, X_{t-2}, \ldots, X_{t-r+1}) \in \Omega^r$ to obtain a Markov process in the usual sense in an enlarged space. Therefore, it will not be necessary to discuss the case of conditional probabilities with a higher degree of memory.

The transition probability W_{ij} can be thought of as a **stochastic matrix**, called the **transition matrix**, that is, it takes non-negative values, $W_{ij} \geq 0$, and satisfies the normalization condition

$$\sum_j W_{ij} = 1 \qquad \forall i \in \Omega \tag{5.1}$$

while from the definition of conditional probability we have

$$p_j(t+1) = \sum_i W_{ij} p_i(t) \tag{5.2}$$

which we can also write in compact vector form as $p(t+1) = p(t) \cdot W$. By iterating (5.2) we obtain

$$p_j(t) = \sum_i W^t_{ij} p_i(0) \tag{5.3}$$

where W^t is the matrix defined by the relation $W^t_{ij} = \sum_k W^{t-1}_{ik} W_{kj}$ with obviously $W^1_{ij} = W_{ij}$. It is easy to see that W^2 is still a stochastic matrix and therefore so is the generic W^t (exercise left to the reader).

Generalizing the previous relation, we have the **Chapman-Kolmogorov equation**

$$W^{t+s}_{ij} = \sum_k W^t_{ik} W^s_{kj} \tag{5.4}$$

which plays a central role in the study of Markov processes both in the discrete case and, with appropriate generalization, in the continuous case. The two Eqs. (5.2) and (5.4) are the equations at the basis of the study of Markov chains.

5.1.1 The Stationary Probability Distribution

A natural and important question is how $p_i(t)$ behaves at long times, that is, whether there exists (and what it is) an **asymptotic stationary distribution** $\pi_i = \lim_{t\to\infty} p_i(t)$ that is a solution of (5.2), i.e., with $\pi = \pi \cdot W$.[3] The main results of the theory of Markov processes concern precisely the existence and uniqueness of the stationary distribution and its determination.

[3] Note that there is always a solution to the equation $\pi = \pi \cdot W$, however there are cases (see below) in which there are multiple stationary solutions.

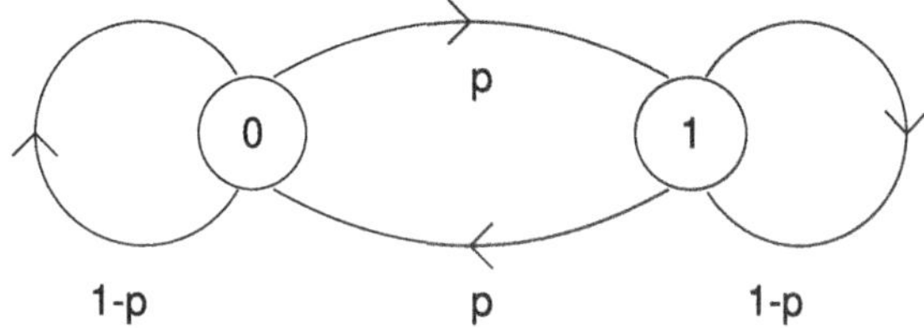

Fig. 5.1 Two-state Markov chain $X = 0, 1$ with transition probabilities $W_{01} = W_{10} = p$ and $W_{00} = W_{11} = 1 - p$

First Example: A Two-State Process

As a first example, let us discuss the case with only two states $\Omega = \{0, 1\}$ in which X_t changes state with probability p and remains in the same state with probability $1 - p$ (Fig. 5.1). We denote by $\varphi(t) = p_0(t)$ the probability that $X_t = 0$ at time t, and thus $p_1(t) = 1 - \varphi(t)$. From (5.2) we have

$$1 - \varphi(t + 1) = p\varphi(t) + (1 - p)(1 - \varphi(t)) = (1 - p) + (2p - 1)\varphi(t)$$

which has the solution (with initial condition $\varphi(0) = \varphi_0$)

$$\varphi(t) = \frac{1}{2} + (1 - 2p)^t \left(\varphi_0 - \frac{1}{2}\right) \tag{5.5}$$

From (5.5) we thus see that for $t \to \infty$, $\varphi(t) \to \pi_0 = 1/2$, independent of the value of φ_0 and p. Note that the convergence is exponentially fast with a characteristic time $\tau = -1/\ln|2p - 1|$.

Second Example: Asymmetric Random Walk on an n-Cycle

Let us now consider a slightly more complicated example in which the random variable can take n integer values $X_t = j$ in $\mathbb{Z}_n = 0, 1, \ldots, n - 1$. We can think of the process as a random walk on the circle in which the point X_t at each step jumps by an angle $\pm 2\pi/n$ with probabilities p and $1 - p$, respectively (see Fig. 5.2). The transition matrix will thus be $W_{i,i+1} = p$, $W_{i,i-1} = 1 - p$, and $W_{ij} = 0$ otherwise

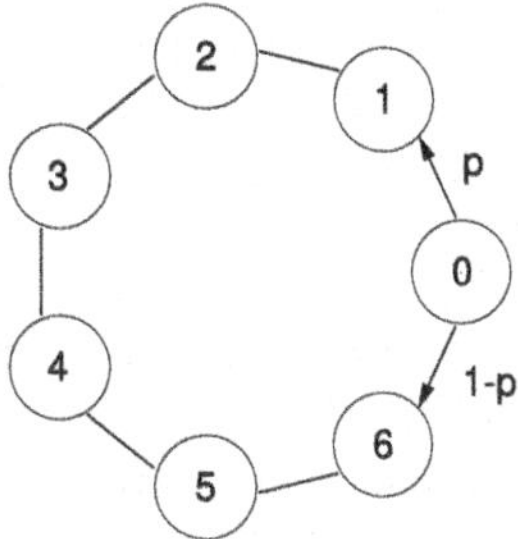

Fig. 5.2 Asymmetric random walk on $\mathbb{Z}_7$. The probability of moving counterclockwise is p, clockwise is $1 - p$

(for the case $n > 2$, obviously). This example is just a particular case of an important class of Markov chains given by a *random walk on a graph* (that is, where the grain can jump, with different probabilities, to a variable number of sites).

It is an elementary exercise to verify that the uniform distribution $\pi_j = 1/n$ is a stationary solution of (5.2). In fact, we have $\sum_i W_{ij}\pi_i = p\pi_{i-1} + (1-p)\pi_{i+1} = 1/n$, independently of the value of p. This result is quite intuitive: after many jumps, the probability of finding the "grain" is the same at each site (and thus no longer depends on the starting site). Beyond the result itself, this problem illustrates well that an important aspect in the study of Markov chains, in addition to determining the asymptotic distribution, is to estimate after how much time the system "forgets" its initial condition.

In this regard, it is interesting to consider the case of even n. As an example, let us take $n = 4$ and $p = 1/2$ and the initial condition $\mathbf{P}(0) = (1, 0, 0, 0)$ (i.e., $X_0 = 0$ with certainty). At time $t = 1$ the probabilities of finding X on one of the n sites will be $\mathbf{P}(1) = (0, 1/2, 0, 1/2)$ and similarly we will have $\mathbf{P}(2) = (1/2, 0, 1/2, 0)$. It is easy to see that we will always have $\mathbf{P}(2k+1) = (0, 1/2, 0, 1/2)$ and $\mathbf{P}(2k) = (1/2, 0, 1/2, 0)$ and thus for $t \to \infty$ $\mathbf{P}(t)$ cannot converge to an invariant distribution. This result holds in general for even n, while in the case of odd n the probability distribution forgets its initial condition and converges to $\pi_j = 1/n$.

5.1.2 *The Role of Barriers: The Gambler's Ruin*

We now introduce, by means of a famous example, the important concept of a barrier in a Markov process.

Suppose we play roulette at the Casino, betting one dollar at a time on either red or black. If we win (with probability p) we gain another dollar, otherwise we lose the bet. We can describe this process with a Markov chain in which the initial state $X_0 \in \mathbb{N}$ is the initial capital and at each step we can jump up or down by a single dollar, exactly as in the random walk example.

Unfortunately, the random walk does not take into account the fact that we can run out of money, that is, X_t cannot take negative values and if at a certain instant we have $X_t = 0$ the subsequent states will always be $X = 0$. The Markov chain corresponding to this situation is called a random walk with an **absorbing barrier** at 0 and is formalized by a transition matrix with $W_{00} = 1$ and $W_{0i} = 0$ for $i > 0$.

Similarly, we can assume that if instead we win a lot, we will be smart enough to leave the Casino. Let's say C is the capital at which we stop playing (or are forced to stop, if C is sufficiently high). In this case, the number of possible states is finite (and equals $C + 1$) and at $X = C$ we will have a second absorbing barrier, that is, $W_{CC} = 1$.

Other types of barriers are possible in Markov chains. For example, suppose that if we lose all our money ($X = 0$) someone lends us a dollar so that we can continue playing. In this case, we have a **reflecting** barrier with $W_{01} = 1$. Other, more complex types of barriers are possible, for example an *elastic* barrier when the random walk reflects only with a certain probability.

An obviously relevant problem for the player is to calculate the probability of bankruptcy, that is, the probability of reaching state 0 before reaching C (and leaving the Casino) starting from a certain capital j. In other words, we want to calculate the probability of reaching a barrier starting from a given initial state. Let F be the event of failure (i.e., the random walk reaches 0 before reaching C); we want to calculate the probability $u_j = P(F|X_0 = j)$. Obviously $u_0 = 1$ (we are already ruined at the start) and $u_C = 0$. For the other cases we can write

$$\begin{aligned} P(F|X_0 = j) &= P(F|(X_1 = j-1) \cap (X_0 = j))P(X_1 = j-1|X_0 = j) \\ &\quad + P(F|(X_1 = j+1) \cap (X_0 = j))P(X_1 = j+1|X_0 = j) \\ &= qP(F|X_1 = j-1) + pP(F|X_1 = j+1) \end{aligned} \tag{5.6}$$

where we have used Bayes' formula[4] and in the last step we have made explicit use of the Markov property and where $q = 1 - p$.

Note now that in the definition of the event F there is no mention of *how much time* it takes to go bankrupt, therefore $P(F|X_1 = j) = P(F|X_0 = j) = u_j$ and thus (5.6) can be written as

$$u_j = qu_{j-1} + pu_{j+1} \tag{5.7}$$

which is a finite difference equation, which can be seen as the discrete version of a differential equation, with boundary conditions $u_0 = 1$ and $u_C = 0$. The solution of (5.7) can be sought in the form $u_j = \alpha^j$ (analogous to the solution $u(x) = e^{\alpha x}$ for a linear differential equation) where the value of the constant α is determined by substituting directly into (5.7)

$$\alpha^j = q\alpha^{j-1} + p\alpha^{j+1}. \tag{5.8}$$

Equation (5.8) admits two solutions: $\alpha = 1$ and $\alpha = q/p$ (note that if $q = p = 1/2$ the two solutions coincide). Analogous to the case of linear differential equations, the general solution of (5.7) will be given by the linear combination of the solutions. If $p \neq q$ the solution sought is therefore

$$u_j = C_1(q/p)^j + C_2,$$

while if $p = q$ we must use as elementary functions α^j and $j\alpha^j$ and therefore

$$u_j = C_1 + C_2 j.$$

[4] If $B = \cup_{i=1}^N B_i$ with $B_i \cap B_j = \emptyset$ if $i \neq j$, using Bayes' formula twice it can be shown that $P(A|B) = \sum_n P(A|B_n)P(B_n|B)$.

We now use the boundary conditions to fix the constants C_1 and C_2. In the first case with $p \neq q$ an elementary calculation leads to $C_1 = 1/(1-(q/p)^C)$ and $C_2 = -(q/p)^C/(1-(q/p)^C)$ and thus

$$u_j = \frac{(q/p)^j - (q/p)^C}{1-(q/p)^C} \qquad (p \neq q), \tag{5.9}$$

while in the second case $C_1 = 1$, $C_2 = -1/C$ and therefore

$$u_j = 1 - j/C \qquad (p = q). \tag{5.10}$$

The second case corresponds to a fair game for which the result (5.10) confirms our intuition: given the symmetry of the problem, if I start with $j > C/2$ I have a higher probability of winning at the Casino than of going bankrupt, and vice versa.

On the contrary, in the case of an unfair game (5.9), we can have a "paradoxical" situation in which the probability of losing is almost certain even for small values of unfairness (these "paradoxes" are the basis of the success of gambling houses). Let us consider, for example, betting on red or black at roulette, for which, due to the presence of green, we have a probability of winning $p = 18/37 < 1/2$ ($18/38$ in American roulette with double zero). Suppose we have an initial capital $j = X_0 = 500$ dollars and aim for a win of $C = 1000$ dollars. The probability of going broke by betting only 1 dollar at a time is, according to (5.9), $P(F|X_0 = 500) \simeq 1 - (p/q)^{C-j} = 1 - 10^{-12}$, that is, we have a probability of about one in 10^{12} of winning, despite the "small" unfairness of the game.

Clearly, this is not the best strategy to make money at the Casino (if there is one at all): it is in fact much more sensible to bet the 500 dollars all at once, for which the probability of losing is only 19/37. We could say that by betting one dollar at a time (and thus being "sure" to lose) we lose the initial stake for the pleasure of continuing to play for a long time. In fact, the average playing time will be much greater than X_0, since this is a random walk, and during this time the player will find themselves many times (we will see how to estimate how many) with a sum significantly greater than X_0.

To conclude, let us consider the case in which $C \to \infty$ (an unrealistic limit, even in Scrooge McDuck's world, but interesting). If the game were fair we would have $u_j = 1 - j/C \to 1$ and thus we would surely lose. Obviously, if the game favors the house, $p < q$, we also have $u_j = 1$, while the interesting case is when $p > q$ (game in favor of the player): in this case $u_j \to (q/p)^j$ and thus there is a finite probability that the game continues indefinitely (but at the price of never leaving the Casino).

5.2 Properties of Markov Chains

An important property of a Markov chain is given by the possibility of reaching a given state j starting from a state i. We will say that state j is **accessible** from state i if there exists a time t such that $W^t_{ij} > 0$. If all states are accessible from all other states in the chain, we will say that the chain is connected or **irreducible**.

A fundamental concept we want to introduce is that of recurrence. Given a state X_t of the process, we ask ourselves whether this state will be visited again or not. The state X is called **recurrent** if the probability of return (at some future time) is 1, while it is called **transient** if there is a finite probability of never returning. Note that a recurrent state is visited infinitely many times, since we can reapply the definition of recurrence every time it is visited. In Fig. 5.3 we show 3 different types of Markov chains with qualitatively different behaviors

To clarify the properties of recurrence, let us reconsider the examples discussed previously. In the case of a random walk with absorbing barriers, the barriers are obviously recurrent while the internal states are all transient (we know for sure that we will end up on one of the barriers). If instead both barriers are reflecting, all (finite) states of the system become recurrent. Similarly for the random walk on an n-cycle (with n odd).

These examples are all made with finite Markov chains, for which there must obviously exist at least one recurrent state (in fact, we have infinite time to visit a finite number of states). If instead we consider infinite chains, we can have the interesting situation in which all states become transient.

Let us consider, for example, a non-symmetric random walk, for instance with $p > q$ as in the case of the game in favor of the player. As we saw with the ruin problem, for $C \to \infty$ there is a finite probability $1 - (q/p)^j$ that starting from state $X_0 = j$ we never reach the origin. If now we remove the barrier at 0 (thus considering $\Omega = \mathbb{Z}$) and start from $X_0 = 0$, with probability p we have $X_1 = 1$ and from here (where we can use the previous result obtained with the barrier, since this is not felt except in state 0) with probability $1 - q/p$ we will never return to 0. Therefore, state 0 never returns with probability $p(1 - q/p) = p - q > 0$ (note

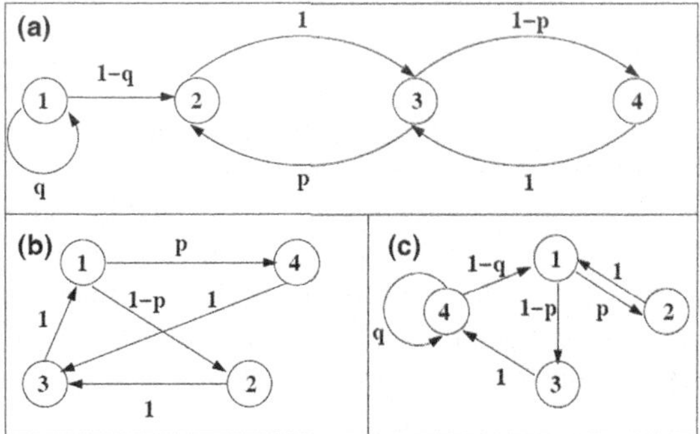

Fig. 5.3 Three examples of Markov chains with 4 states. (**a**) Reducible chain with state 1 transient and states 2, 3 and 4 recurrent and periodic with period 2. (**b**) Irreducible periodic chain with period 3. (**c**) Irreducible ergodic chain

that if $p = 1$ the random walk advances deterministically along the positive axis and trivially never returns to the origin). Since now 0 is an arbitrary point in Ω, the same result holds for all points and therefore all states of the asymmetric random walk without barriers are transient.

The asymmetry of the random walk ($p \neq q$) is fundamental: if we consider the symmetric case, the gambler's problem has taught us that starting from any state $j > 0$ we reach 0 with certainty. This result also generalizes to any pair of points in Ω and thus every state is persistent and is visited infinitely often. Finally, let us recall that the property of persistence also depends on the dimensionality of the problem. The infinite random walk (without barriers and symmetric) is persistent (at every point) also in two dimensions, but is no longer so in $d \geq 3$.

The last property of Markov chains we want to recall is that of periodicity. A chain is called **periodic** if the possible return times to a state i are multiples of a period $t_i > 1$, that is, $W_{ii}^{t_i} > 0$. As an example, let us again consider the random walk on an n-cycle discussed earlier in Sect. 5.1.1. As shown, if n is even the successive states of X_t will belong to two disjoint sets (even and odd) and therefore the period of the chain is $t = 2$. Instead, if n is odd the chain is not periodic. Now consider the case of even n and suppose we start from a state $X_0 = j$ even. After an even number of steps, the distribution of the chain $p_i(2t)$ will have support only on even states i, while after an odd number of steps $p_i(2t+1)$ will have support only on odd states. It is therefore evident that the distribution does not converge as $t \to \infty$. This result holds in general and shows that for periodic chains $p_i(t)$ cannot tend toward an invariant probability π_i.

5.2.1 *Ergodic Markov Chains*

Let us now discuss an important class of Markov chains called **ergodic**, that is, *chains with a finite number of states, connected, aperiodic, and with all states persistent*. A necessary and sufficient condition for a Markov chain to be ergodic is that for some $t \geq 1$ we have $W_{ij}^t > 0$ for every pair $i, j \in \Omega$.

The first important property of ergodic Markov chains is that the stationary (or invariant) distribution π_j is unique and is obtained at arbitrarily long times from the evolution of a generic initial distribution $p_j(0)$. Recall that the invariant distribution is a solution of $\pi = \pi \cdot W$, which is equivalent to requiring that π is an eigenvector of the transition matrix W with eigenvalue 1 (and obviously that the eigenvector has non-negative components that sum to unity).

The value π_j represents the asymptotic probability that the state takes the value $X = j$ and, if the chain is ergodic, it also represents the fraction of time in which $X = j$. Since all states are persistent, it is interesting to ask how long it takes for the initial condition $X_0 = j$ to return to state j, that is, what is the value of the **first return time** (or recurrence time) T_j.

For ergodic chains, the following result holds, which finds important applications (for example, in the Monte Carlo method). Let A be any function that takes the value A_i in the generic state $i \in \Omega$ and let $\{i_0, i_1, \ldots, i_{t-1}\}$ be a realization of the Markov process. For almost all realizations, we have

$$\lim_{t\to\infty} \frac{1}{t} \sum_{k=0}^{t-1} A_{i_k} = \sum_{i\in\Omega} A_i \pi_i . \tag{5.11}$$

The meaning of ergodicity becomes transparent if we take as observable the function that is $A_i = 1$ if $i = r$ while $A_i = 0$ if $i \neq r$, then (5.11) becomes

$$\lim_{t\to\infty} \frac{1}{t} \sum_{k=0}^{t-1} A_{i_k} = \pi_r$$

in other words, the fraction of time spent in state r asymptotically (for large t) coincides with the probability of the state itself π_r.

Let us also note that the concept of ergodicity, which will be revisited in Chap. 7 when we deal with chaotic systems, is an extension, for non-independent variables, of the law of large numbers discussed in Chap. 3.

We now want to study the statistical properties of the first return times and in particular the mean value. The *return time theorem* or *Kac's lemma*, a fundamental result for ergodic systems, relates the mean value of the return times $E(T_j)$ to the stationary distribution according to the relation:

$$E(T_j) = 1/\pi_j \tag{5.12}$$

We observe that (5.12) expresses the intuitive, but nontrivial, fact that the average time between two visits to state j is the inverse of the time spent in state j (which, as seen, is given by π_j).

Perhaps the simplest way to prove (5.12) is to exploit ergodicity. Let us consider a realization of the process $\{i_0, i_1, \ldots\}$ with initial state $i_0 = j$ and denote by $T_j^{(1)}$ the time the system takes to return for the first time to j, $T_j^{(2)}$ the time for the second return (starting from $T_j^{(1)}$) and so on. The total time the system takes to return N times to j will therefore be

$$t_N = \sum_{k=1}^{N} T_j^{(k)} .$$

During this time, the system will have spent in state j the fraction of time

$$f_N = \frac{N}{t_N} .$$

Since the system is ergodic, we have $\lim_{N\to\infty} f_N = \pi_j$ and therefore we can calculate

$$E(T_j) = \lim_{n\to\infty} \frac{1}{N} \sum_{k=1}^{N} T_j^{(k)} = \lim_{N\to\infty} \frac{1}{f_N} = \frac{1}{\pi_j}.$$

A further interesting result that holds for ergodic Markov chains is the exponentially fast convergence of $p_i(t)$ to the stationary distribution π_i:

$$p_t = \pi + o\left(e^{-t/\tau_c}\right) \tag{5.13}$$

where the characteristic time τ_c can be explicitly calculated from the transition matrix W. By definition, the stationary distribution, which obeys $\pi = \pi \cdot W$, is an eigenvector of W with eigenvalue 1. It is possible to show (it is a theorem of linear algebra) that the eigenvalues α_k of W are all in modulus less than 1, $|\alpha_k| < 1$, except for the one associated with π. Ordering the eigenvalues in non-increasing order, $1 = \alpha_1 > |\alpha_2| \geq |\alpha_3| \ldots$ the first eigenvalue is non-degenerate, that is $|\alpha_2| < 1$, and this ensures (5.13) with $\tau_c = -1/\ln|\alpha_2|$.

5.2.2 *Reversible Markov Chains*

An important class of Markov chains, which finds many physical applications, is that of **reversible** chains. Intuitively, a Markov chain is reversible if the probability of any sequence $(X_0, X_1, \ldots X_t)$ is the same as that of the mirror sequence $(X_t, X_{t-1}, \ldots X_0)$. A necessary and sufficient condition for reversibility is that the transition probabilities satisfy the **detailed balance** condition

$$\pi_i W_{ij} = \pi_j W_{ji} \tag{5.14}$$

for all $i, j \in \Omega$ (note that in the definition no sum is implied).

Recall the important result that *any Markov chain that satisfies detailed balance is ergodic with invariant probability given by the π_i solution of (5.14)*. It is indeed possible to verify that the condition for ergodicity is satisfied if (5.14) holds. Moreover, summing both sides of (5.14) over i and recalling the normalization condition (5.1) we have

$$\sum_{i\in\Omega} \pi_i W_{ij} = \sum_{i\in\Omega} \pi_j W_{ji} = \pi_j$$

which is the stationarity condition for the probability. The verification of the detailed balance condition, when it holds, is one of the simplest methods to verify the ergodicity of the Markov chain.

As mentioned, detailed balance implies the reversibility of the Markov process, in the sense that the probability of a given sequence $(i_0, i_1, \ldots, i_n)$ is the same as that of the mirror sequence $(i_n, i_{n-1}, \ldots, i_0)$. Indeed, by repeatedly applying (5.14) to the sequence

$$\begin{aligned} \pi_{i_0} W_{i_0 i_1} W_{i_1 i_2} \ldots W_{i_{n-1} i_n} &= W_{i_1 i_0} \pi_{i_1} W_{i_1 i_2} \ldots \\ \ldots W_{i_{n-1} i_n} &= W_{i_1 i_0} W_{i_2 i_1} \ldots W_{i_n i_{n-1}} \pi_{i_n} \end{aligned} \tag{5.15}$$

Translated in terms of probability, Eq. (5.15) tells us that, starting from the stationary distribution,

$$P(X_0 = i_0, X_1 = i_1, \ldots X_n = i_n) = P(X_0 = i_n, X_1 = i_{n-1}, \ldots X_n = i_0)$$

and therefore the probability of a time sequence is equal to that of the reverse sequence.

As an example, let us again consider the asymmetric random walk on an n-cycle from Sect. 5.1.1. The stationary distribution is uniform, $\pi(j) = 1/n$, but we have

$$\pi(j)P(j, j+1) = \frac{p}{n} \neq \frac{1-p}{n} = \pi(j+1)P(j+1, j)$$

unless $p = 1 - p = 1/2$. Therefore, the asymmetric random walk does not satisfy detailed balance and is not reversible (as is obvious, since it has a preferred direction of rotation).

5.2.3 *The Ehrenfest Model for Diffusion*

In 1907, Paul and Tatiana Ehrenfest, while discussing the mathematical and conceptual foundations of statistical mechanics, proposed a simple Markovian model to clarify the statistical interpretation of the second law of thermodynamics. The model, although mathematically simple, is very instructive and allows one to unambiguously understand some apparently paradoxical aspects of nonequilibrium statistical mechanics. Let us consider N molecules in a container divided into two parts (let us call them A and B) by a permeable partition. The state of the system at time t is described by the value $k(t)$ of molecules in sector A (and thus $N - k$ is the number of molecules in B). In the dynamics proposed by Ehrenfest, at each (discrete) time step, one of the N molecules is chosen at random and moved to the other sector.

The Ehrenfest model is clearly a Markov process with a transition matrix W that takes nonzero values only on the lines adjacent to the diagonal. In fact, we have

$$\begin{aligned} W_{k,k-1} &= k/N \\ W_{k,k+1} &= (N-k)/N \end{aligned} \tag{5.16}$$

while $W_{ij} = 0$ if $|i-j| \neq 1$.

Comparing with the transition matrix of the random walk described in the previous chapter, we see that the Ehrenfest model can be rephrased as a random walk in which the probability of jumping to the left ($k \to k-1$) or to the right ($k \to k+1$) depends on the position of the grain. In particular, Eq. (5.16) tells us that the probability of jumping to the left is greater than that of jumping to the right for $k > N/2$ (and vice versa). Therefore, the model represents a random walk in the presence of an "attractive force" toward the point $k = N/2$. This consideration allows us to conjecture that, unlike the case of a pure random walk, the stationary distribution for the Ehrenfest model will be concentrated around $k = N/2$.

The didactic aspect of the Ehrenfest model is that the stationary distribution π_k (the probability of having k molecules in sector A at long times) can be obtained exactly in an elementary way. The method, which we report below, can be applied to a generic transition matrix that has nonzero values only on the diagonal and on the adjacent lines. From the condition for the stationary distribution, we can explicitly write

$$\begin{aligned} \pi_0 &= W_{1,0}\pi_1 \\ \pi_1 &= W_{0,1}\pi_0 + W_{2,1}\pi_2 \\ \pi_2 &= W_{1,2}\pi_1 + W_{3,2}\pi_3 \end{aligned}$$

and so on. This system of equations can be solved iteratively starting from the first equation, keeping π_0 as a free parameter that will be fixed by normalization. We then have, using (5.16),

$$\begin{aligned} \pi_1 &= N\pi_0 \\ \pi_2 &= \frac{N(N-1)}{2}\pi_0 \\ \pi_3 &= \frac{N(N-1)(N-2)}{3 \cdot 2}\pi_0 \end{aligned} \tag{5.17}$$

and therefore in general $\pi_k = \pi_0 N!/[k!(N-k)!]$. The normalization condition requires $\pi_0 = 1/2^N$ so that

$$\pi_k = \frac{1}{2^N}\binom{N}{k} \tag{5.18}$$

that is, the stationary distribution is binomial. This means that after many steps, the distribution is the same as would be obtained by putting each molecule in sector A or B with equal probability $1/2$: under Ehrenfest dynamics, the system relaxes to the binomial distribution independently of the initial state. Note that, if $N \gg 1$, the distribution (5.18) is strongly peaked around the most probable value $k = N/2$, where therefore the system spends most of the time. Even with $N = 10^6$ molecules (a "small" number from a thermodynamic point of view), the probability of observing a fluctuation of 1% (that is, 505,000 molecules in A) is on the order of 10^{-23}. In this sense, the dynamics of the Ehrenfest model, viewed from a macroscopic perspective with large N, is irreversible: once the equilibrium state is reached, the system will not spontaneously deviate from it.

To avoid misunderstandings, let us note that the Ehrenfest model is governed by a formally reversible process, in the sense that it satisfies detailed balance (5.6), as can be seen directly by using (5.16) and (5.18). The fact that in the expansion into an empty sector (i.e., with $k(0) = 0$) the direction of time is evident is a consequence of the particular initial condition. Starting from the "typical" equilibrium initial condition ($k(0) = N/2$), it is clear that we can reverse the arrow of time and obtain an equally plausible process, as shown in Fig. 5.4.

It may be interesting to apply the results of the previous section on return times to the Ehrenfest model. We have seen that the mean first return time to state k, $E(T_k)$, is given by $1/\pi_k$. For the most probable state, $k = N/2$, we obtain, using the Gaussian approximation for the binomial distribution, $E(T_{N/2}) \simeq \sqrt{\pi N/2}$. For a macroscopic example with $N \sim 10^{23}$, we get $E(T_{N/2}) = O(10^{11})$, a time that is only apparently large because the time unit used is the jump time of a single molecule.

If instead we consider the classic experiment of expansion into a vacuum, for which the initial state is given by $k(0) = N$, the return time (that is, the spontaneous return of all molecules to sector A) is $E(T_0) = 1/\pi_0 = 2^N \simeq 10^{10^{22}}$, a physically enormous time regardless of the time unit used. Therefore, the irreversibility of the expansion into an empty sector is due to the particular initial condition chosen.

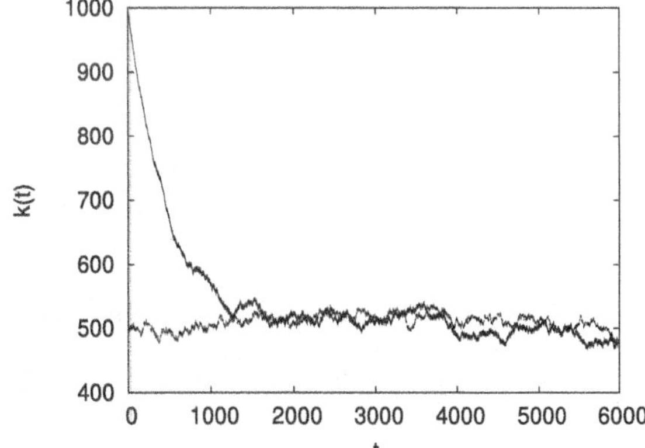

Fig. 5.4 Two examples of the evolution of the Ehrenfest model with $N = 1000$ and initial conditions $k(0) = N$ (upper solid line) and $k(0) = N/2$ (dashed line). The (apparent) irreversibility of the process starting from $k(0) = N$ is evident, while the process with initial condition $k(0) = N/2$ shows that the process is reversible

The simplicity of the Ehrenfest model also allows us to explicitly calculate how the system relaxes to the most probable state $k = N/2$ starting from a state far from equilibrium. Let us consider the time evolution of the model starting from an arbitrary initial condition $k(0)$:

$$k(t+1) = k(t) + \Delta(k(t)) \tag{5.19}$$

where $\Delta(k) = +1$ with probability $W_{k,k+1}$ and $\Delta(k) = -1$ with probability $W_{k,k-1}$ given by (5.16). Obviously, if $k(t)$ is different from $N/2$, we expect that on average $k(t+1)$ will approach the equilibrium value. In fact, we can write, from (5.19),

$$\langle k(t+1)\rangle = \langle k(t)\rangle + W_{k,k+1} - W_{k,k-1} \tag{5.20}$$

Using the transition probabilities (5.16), (5.20) can be written as

$$\left[\langle k(t+1)\rangle - \frac{N}{2}\right] = \left(1 - \frac{2}{N}\right)\left[\langle k(t)\rangle - \frac{N}{2}\right]$$

which, iterated in time, gives us the evolution for $\langle k(t)\rangle$

$$\left[\langle k(t)\rangle - \frac{N}{2}\right] = \left(1 - \frac{2}{N}\right)^t \left[\langle k(0)\rangle - \frac{N}{2}.\right] \tag{5.21}$$

Therefore, considering an initial state $k(0)$ far from equilibrium, for example $k(0) \simeq N$, the mean number of molecules in sector A decreases exponentially toward the stationary value $N/2$. Similarly, it is also possible to calculate the variance $\sigma^2(t) = \langle k^2(t)\rangle - \langle k(t)\rangle^2$. If the initial condition $k(0)$ is known with certainty, then $\sigma^2(0) = 0$ and it is found that $\sigma^2(t)$ increases monotonically until it tends to $\lim_{t\to\infty} \sigma^2(t) \propto N$.

We can recognize in (5.21) a particular form of Boltzmann's H theorem for the function $H(t) = \langle k(t)\rangle - N/2$. The H theorem, derived by Boltzmann for a gas of molecules obeying the laws of mechanics, is a central pillar of the microscopic interpretation of the irreversibility of thermodynamics. (entropy increase law). It is well known that Boltzmann had serious difficulties getting his theorem accepted, difficulties that were formalized in particular in two "paradoxes." The recurrence paradox, discussed earlier, which is "resolved" by considering the fact that the return time is, from a physical point of view, infinite; and the reversibility paradox: given a microscopic dynamics that leads to a decrease in the function $H(t)$, it is enough to reverse the velocities of the molecules at a certain point to obtain an increasing H function.

The reversibility paradox can be rephrased in the Ehrenfest model by stating that, given a realization of the Markov chain that brings $k(0)$ towards $N/2$, there will be the inverse realization, microscopically equally probable, which starting from the final state $k(t)$, will take it away from the equilibrium value. The explanation of the apparent paradox is that the function $H(t)$ is defined by an averaging operation and that among all possible realizations of the Markov chain, there is an extremely larger number for which $H(t)$ decreases.

Let us again consider an initial condition far from equilibrium $k(0) \simeq N$. By definition, $k(t)$ will fluctuate with an amplitude $\sigma(t)$ around the mean value $\langle k(t)\rangle$. As long as $k(t) \gg N/2$ we will have $\sigma(t) \sim \sqrt{N}$ and therefore $k(t) - N/2$ will follow, apart from very small fluctuations, the decreasing trend described by (5.21). Asymptotically, $k(t)$ will then stabilize around the value $N/2$ with excursions of the order of $\sqrt{N}$.

5.3 How to Use Markov Chains for Practical Purposes: The Monte Carlo Method

Let us consider the following problem: a random process can assume N different states labeled by the integers $j = 1, 2, \ldots, N$, with probabilities $p_1, p_2, \ldots, p_N$. We want to determine the mean value of some observables:

$$< A >= \sum_{j=1}^{N} A_j p_j \tag{5.22}$$

where A_j is the value in state j. For example, in statistical mechanics, if the energy states are discrete (as in spin systems), denoting by E_j the energy of state j we have

$$p_j = \frac{1}{Z} e^{-\beta E_j}$$

where Z is the partition function: $Z = \sum_j e^{-\beta E_j}$, and $< E >$ is the internal energy.

The problem is interesting, and not trivial, in the case where $N \gg 1$ and therefore it is not possible to explicitly perform the sum in (5.22). As an example, consider a system of M spins with two possible states (as in the Ising model), then the total number of possible configurations is 2^M, an enormous number even if M is very far from Avogadro's number: if $M = 100$ (a very small two-dimensional system, only 10×10) we have $N = 2^{100} \simeq 10^{30}$, a decidedly large number.[5]

In many situations, for example in the case just mentioned, even if the number of states is enormous, the vast majority of them are inessential in the calculation of

[5] In the case where there is degeneracy, one can write (5.22) in the form

$$< A >= \sum_{j=1}^{N^*} g_j A_j p_j$$

with $N^* < N$, but in general this does not simplify the calculation much.

the average (5.22) since they have negligible probability. Obviously, this does not help automatically, in fact if one cannot "skip" the states with negligible probability without first checking that p_j is small, it is just as well to sum the term $A_j p_j$. Is it possible to calculate the mean value (5.22), at least approximately, in these cases? The Monte Carlo method is a way to tackle the problem.

The idea is as follows: given the probabilities $\{p_j\}$, one finds an ergodic Markov chain (i.e., a suitable transition matrix $W_{i,j}$) such that the $\{p_j\}$ are the invariant probabilities of the Markov chain. At this point, exploiting ergodicity, the mean value (5.22) is determined by a time average along a sufficiently long "trajectory" $\{i_1, i_2, \ldots, i_T\}$ of the process described by the Markov chain:

$$\overline{A}^T = \frac{1}{T}\sum_{t=1}^{T} A_{i_t} \to < A >= \sum_{j=1}^{N} A_j p_j. \tag{5.23}$$

The trajectory is determined by the following procedure: once the state i_t is identified, the state i_{t+1} is the result of rolling a "loaded die" that takes values $j = 1, 2, \ldots$ with probability $W_{i_t,j}$. Obviously, the "loaded die" in numerical calculation is a random number generator.[6] From a strictly mathematical point of view, Eq. (5.23) is a consequence of ergodicity. This naturally raises the question: why does the Monte Carlo method work in practice even when $T \ll N$?

Let us denote by $f_k(T)$ the frequency of state k in the trajectory $\{i_1, i_2, \ldots, i_T\}$; we have

$$\frac{1}{T}\sum_{t=1}^{T} A_{i_t} = \sum_{k} A_k f_k(T)$$

while the exact value of $< A >$ is given by two contributions

$$< A >= \sum_{k \in \mathcal{S}_1} A_k p_k + \sum_{j \in \mathcal{S}_2} A_j p_j$$

where $\mathcal{S}_1$ denotes the set of states with non-negligible probability, i.e., $p_k > \epsilon$, and $\mathcal{S}_2$ the set of states with $p_k < \epsilon$. The time average approximates the first contribution; in fact, for finite T we have $f_k(T) \simeq p_k$ if T is at least $O(1/p_k)$, whereas if T is small compared to $1/p_j$, then state j will practically not be visited by the trajectory of length T.

Thus, the Monte Carlo method "automatically selects" the states with probability greater than $\epsilon \sim O(1/T)$, avoiding "wasting time" visiting those with small probability.

[6] A random number X is generated uniformly distributed between 0 and 1, the state at time $t+1$ will be 1 if $0 \le X < W_{i_t,1}$, or the integer j (greater than 1) that satisfies the condition $W_{i_t,j-1} \le X < W_{i_t,j}$.

Obviously, in the Monte Carlo method there are many practical issues, as in all numerical methods, the first of which is the estimation of the error in the calculation of the average.[7]

Let us note that, given the probabilities $\{p_j\}$, there are several possible choices for the transition matrix $W_{i,j}$ such that the Markov chain is ergodic with invariant probability $\{p_j\}$. How should one decide on the choice?

In the statistical mechanics of lattice systems, one generally uses an algorithm called Metropolis[8] (also called Metropolis-Hastings) which satisfies detailed balance, and is constructed following a physical intuition. The "recipe" is as follows. Given the invariant probabilities $\{p_j\}$,

(a) a graph is constructed whose vertices are the states $\{S_j\}$; a pair of states is connected (by a directed arrow) if the transition $i \to j$ is allowed;
(b) the graph must be connected (so that the chain is irreducible and ergodic);
(c) each state S_i is connected to a few sites (this is a requirement motivated only by computational convenience), we denote by d_i the number of arrows leaving S_i;
(d) suppose we are at time t in state S_i. The state at time $t+1$ is determined by the following rule: one selects (with equal probability $1/d_i$) a state S_j among those connected to S_i; then the new state will be S_j with probability

$$min\{\frac{p_j d_i}{p_i d_j}, 1\}$$

or will remain in S_i with probability

$$1 - min\{\frac{p_j d_i}{p_i d_j}, 1\}\,.$$

This choice, which is easy to use for spin systems on a lattice and, with appropriate modifications, also for interacting particle systems, corresponds to an ergodic Markov chain.

In spin systems with local interaction (such as the Ising model), the practical rule is as follows:

(1) One selects (with equal probability $1/M$) one of the M spins, say the k-th is chosen; one calculates the local contribution to the energy (given by the interaction between the k-th spin and its nearest neighbors), let us denote this value by E;

[7] The main difficulty is due to the fact that the variables $\{A_{i_1}, A_{i_2}, \dots.\}$ are not uncorrelated, so an estimate of the error on the average depends on the characteristic time of the process τ_c (which can be estimated from the behavior of the correlation functions); the error on the average is $O(\sqrt{\tau_c/T})$.

[8] N. Metropolis, A. Rosenbluth, M. Rosenbluth, M. Teller and E. Teller. *J. Chem. Phys.* **21**, 1087 (1953).

(**2**) the sign of the k-th spin is flipped; $\sigma_k \to -\sigma_k$ and the local contribution to the energy E' of this new configuration is calculated;
(**3**) if $E' < E$ the new value of the k-th spin is accepted;
(**4**) if $E' > E$ a random number X uniformly distributed between 0 and 1 is drawn; if $X < e^{-\beta(E'-E)}$ the flip is accepted, otherwise the old value is kept;
(**5**) return to step (**1**).

Let us note that from a strictly mathematical point of view it is possible to use different algorithms (for example, that of Swensen and Wang). Two observations:

(a) the dynamical properties of the Markov chain used in Monte Carlo do not have an immediate physical meaning; this is evident by observing that different chains can have the same invariant probabilities;
(b) detailed balance (used in the Metropolis algorithm) is not at all, as is sometimes implied, an essential ingredient.

The two previous points can be understood from the following example (absolutely not interesting from a practical point of view). Let us consider the case with $p_j = 1/N$, where N is odd; we can construct a Markov chain with a random walk on a discrete ring with periodic boundary conditions. The two transition matrices

$$(a) \quad W_{i,i\pm1} = \frac{1}{2}$$

$$(b) \quad W_{i,i\pm1} = \frac{1}{2} \pm \Delta \quad 0 < \Delta < \frac{1}{2}$$

have the same invariant probability $p_j = 1/N$, however, their dynamical properties are very different. For case (**a**), in which detailed balance holds and there is no preferred direction, the time T required for a good estimate of the mean value is $O(N^2)$. On the contrary, in the Markov chain (**b**), detailed balance does not hold and there is a preferred direction (in fact, there is an "average current" 2Δ), T is $O(N/\Delta)$.

5.4 Continuous-Time Processes: The Master Equation

We conclude this chapter with a brief mention of the case of discrete-state systems in continuous time, that is, those processes described by a **master equation** as explained in the introduction.

Let us start from the Chapman-Kolmogorov equation (5.4) for a process as a function of a time $t \in \mathbb{R}$ which we write in the form

$$W_{ij}^{t+\Delta t} = \sum_k W_{ik}^t W_{kj}^{\Delta t} \tag{5.24}$$

where we are interested in the limit $\Delta t \to 0$. In this limit, we expect that the probability of not changing state is much greater than the probability of changing it, and therefore we assume that

$$W_{kj}^{\Delta t} = \begin{cases} 1 - \kappa_k \Delta t + O(\Delta t^2) \; if \; k = j \\ T_{kj} \Delta t + O(\Delta t^2) \quad\; if \; k \neq j \end{cases} \tag{5.25}$$

The matrix T_{kj} represents the transition rate from state k to state j per unit time, and from condition (5.1) it must hold that

$$\sum_{j \neq k} T_{kj} = \kappa_k. \tag{5.26}$$

Substituting (5.25) into (5.24) and subtracting W_{ij}^t from both sides, we obtain

$$W_{ij}^{t+\Delta t} - W_{ij}^t = \Delta t \left(\sum_{k \neq j} T_{kj} W_{ik}^t - \kappa_j W_{ij}^t. \right).$$

Dividing by Δt, using (5.26), and taking the limit $\Delta t \to 0$, we obtain the **master equation**

$$\frac{dW_{ij}^t}{dt} = \sum_{k \neq j} T_{kj} W_{ik}^t - \sum_{k \neq j} T_{jk} W_{ij}^t. \tag{5.27}$$

The same equation governs the time evolution of the probability $p_i(t)$. In fact, by differentiating (5.3) with respect to time and using (5.27), we obtain

$$\frac{dp_j(t)}{dt} = \sum_{k \neq j} T_{kj} p_k(t) - \sum_{k \neq j} T_{jk} p_j(t). \tag{5.28}$$

The first term represents the transition rate from all states $k \neq j$ to state j, while the second represents the rate of leaving state j.

5.4.1 An Example: Birth and Death Processes

Let us now consider a simple and classic example of a stochastic process governed by a master equation: the evolution of a population of organisms (for example, a colony of bacteria). We can suppose, as a first approximation, that both the birth rate of new bacteria and the death rate are proportional to the number of individuals

n present at a given time.[9] This is clearly a Markov process and, assuming that at a given instant at most one individual can be born, we will have that the transition rate is nonzero only between neighboring states (that is, it is a so-called *one-step* process)

$$T_{n,n+1} = bn \quad \text{and} \quad T_{n,n-1} = dn$$

where b and d are two constants that depend on the type of population. Equation (5.28) becomes

$$\frac{dp_n(t)}{dt} = (n+1)dp_{n+1}(t) + (n-1)bp_{n-1}(t) - n(b+d)p_n(t).$$

The solution can be found exactly by making use of the generating functions introduced in Chap. 2. Let us consider, for simplicity, a pure mortality process, that is, one for which $b = 0$. Without loss of generality, we can take $d = 1$ (the value of d simply rescales time), and therefore we have

$$\frac{dp_n(t)}{dt} = (n+1)p_{n+1}(t) - np_n(t). \tag{5.29}$$

We multiply both sides by s^n and sum over n to obtain

$$\frac{d}{dt}\sum_{n=0}^{\infty} p_n s^n = \sum_{n=0}^{\infty}(n+1)p_{n+1}s^n - \sum_{n=0}^{\infty} np_n s^n.$$

Recalling the definition of the generating function, $G(s,t) = \sum_{n=0}^{\infty} p_n(t)s^n$, we have

$$\frac{\partial G(s,t)}{\partial t} = (1-s)\frac{\partial G(s,t)}{\partial s}. \tag{5.30}$$

It is straightforward to verify (or derive using the method of characteristics) that, given the new variables

$$\begin{cases} x = (1-s)e^{-t} \\ \tau = t \end{cases}$$

Equation (5.30) becomes $\frac{\partial G}{\partial \tau} = 0$ and therefore the desired generating function will be any function $f(x)$ of the single argument x

$$G(s,t) = f\left((1-s)e^{-t}\right).$$

[9] Obviously, in doing so we neglect any effects due to resource limitation and other important factors.

The form of f is determined by the initial condition. If we assume that at the initial time the population consists of N individuals so that $p_n(0) = \delta_{n,N}$, we have $G(s,0) = s^N = f(1-s)$. Therefore, the desired function will be $f(x) = (1-x)^N$, which, at a generic time t, gives us the solution

$$G(s,t) = \left(1 - (1-s)e^{-t}\right)^N . \tag{5.31}$$

From (5.31) we can explicitly, though laboriously, calculate $p_n(t)$. More simply, we can calculate the moments of the distribution. For example, we immediately obtain for the mean value the unsurprising result

$$\langle n(t) \rangle = \frac{\partial G(s,t)}{\partial s}|_{s=1} = Ne^{-t}.$$

Exercises

5.1 Consider a Markov chain with 2 states with transition probabilities

$$P_{1\to 1} = 1-a,\ P_{1\to 2} = a,\ P_{2\to 1} = b,\ P_{2\to 2} = 1-b$$

with $0 < a < 1$, $0 < b < 1$. Given $(P_1(0), P_2(0))$, explicitly calculate $(P_1(t), P_2(t))$ and check the convergence to the invariant probability. Show that, for this chain, detailed balance always holds.

5.2 A Markov chain with N states whose transition matrix satisfies the property

$$\sum_{j=1}^{N} P_{j\to k} = 1 \quad for\ every\ k$$

is called doubly stochastic. Show that if the chain is doubly stochastic and ergodic, the invariant probabilities are $\pi_j = 1/N$.

5.3 An old Black Forest proverb says: if today the weather is nice, tomorrow it will be nice 2 times out of 3; if today it is bad, tomorrow it will be bad 3 times out of 4. Calculate

(a) how many nice days there are on average in a year;
(b) the probability of having at least 5 nice days in a row after a bad day.

5.4 Consider a very simplified meteorology with only three types of weather: sun, rain, and wind (let us denote them by 1, 2, and 3). The transition rules are as follows: for each state there is a probability 1/2 of not changing and a probability 1/4 of going to one of the other two.

(a) Calculate the invariant probabilities.
(b) Let us group rain and wind together, so we have only two states,

let us denote the old 1 as I and the one obtained by grouping 2 and 3 as II; is this still a Markov process? More precisely: given the original Markov process x_t (with $x_t = 1$ or 2 or 3), is the variable y_t defined as follows $y_t = I$ if $x_t = 1$, $y_t = II$ if $x_t \neq 1$, still a Markov process?

5.5 Consider a Markov chain with 3 states with transition probabilities

$$P_{1\to 2} = P_{3\to 1} = 1, \quad P_{2\to 1} = p, \quad P_{2\to 3} = 1 - p,$$

where $0 < p < 1$, and the other transitions have zero probability. Let us group together states 1 and 2 (we denote the new state by I) and rename the old state 3 as II; show that the new stochastic process $\{y_t\}$ thus defined is not a Markov chain.

5.6 A maze is composed of 4 rooms; in each room there are 3 identical doors, each connecting to the other rooms. Room 4 is a trap (once entered, it is impossible to leave). Each time a bell rings, the mouse changes room by choosing, with equal probability, one of the 3 available doors, without remembering the past. At the initial time, the mouse is in room 1; calculate the probability that after N rings the mouse is not trapped.

5.7 A fair die is rolled n times; let x_n denote the maximum result obtained. Show that the variable x_n is described by a Markov chain.

5.8 Consider a random walk on a lattice of length L. At sites $n = 2, 3, .., L - 1$ there is a probability $1/2$ of moving forward or backward. At $n = L$ one always moves backward with probability 1, while at $n = 1$ one always moves forward with probability 1. Calculate the invariant probabilities.

5.9 Consider a random walk on the non-negative integers with the following transition probabilities: $P_{0\to 1} = 1$ and

$$P_{n\to n+1} = \frac{r}{1+r}, \quad P_{n\to n-1} = \frac{1}{1+r}, \quad \text{if } n \geq 1.$$

Show that if $0 < r < 1$ there always exists an invariant probability; find this probability.

5.10 Consider a random walk on the non-negative integers with the following transition probabilities:

$$P_{n\to n-1} = 1 \text{ if } n \geq 1, \quad P_{0\to n} = f_n,$$

and the others are zero, obviously

$$\sum_{n=0}^{\infty} f_n = 1.$$

Calculate the invariant probabilities in the case where f_n decays to zero sufficiently rapidly, that is, more rapidly than $n^{-\beta}$ with $\beta > 2$.

This process (which seems very artificial) is a stochastic model for an interesting class of one-dimensional chaotic maps; in the case $f_n \sim n^{-\beta}$ with $1 < \beta \leq 2$ there are no invariant probabilities and the process is relevant for "sporadic chaos", see X.-J. Wang "Statistical physics of temporal intermittency" Phys. Rev. A **40**, 6647 (1989).

5.11 Consider a branching process in which, after one generation, an element produces m individuals with probability

$$P_m = \frac{\lambda^m}{m!} e^{-\lambda} \quad m = 0, 1, 2, \ldots$$

Find the probability of having j elements at time $t + 1$ if at time t there are n.

5.12 Consider a Markov chain with 3 states with transition probabilities

$$P_{1\to 2} = P_{1\to 3} = \frac{1}{2}, \; P_{2\to 1} = P_{3\to 1} = 1,$$

and the others are zero.

(a) Show that there does not exist a limiting distribution.
(b) Calculate the fraction of time that the system spends in state $j = 1, 2, 3$ in the infinite time limit.

Recommended Readings

Markov chains are discussed in some previously cited texts (for example, Gnedenko and Renyi). Good introductory books with particular attention to applications:

O. Häggström, *Finite Markov Chains and Algorithmic Applications* (Cambridge University Press, Cambridge, 2002)

A.T. Bharucha-Reid, *Elements of the Theory of Markov Processes and Their Applications* (Dover Publications, New York, 2010)

A good introduction is given by a manuscript (never published) by Gian Carlo Rota and Kenneth Baclawski *An introduction to probability and random processes* which can be obtained from www.freescience.info

On the Ehrenfest model:

P. Ehrenfest, T. Ehrenfest, *The Conceptual Foundation of the Statistical Approach in Mechanics* (Cornell University Press, New York 1956);

M. Kac *Probability and Related Topics in Physical Sciences* (Am. Math. Soc., Providence, 1957)

Chapter 6
Stochastic Processes with Continuous States and Time

Chapter 5 was dedicated to stochastic processes with discrete states, that is, Markov chains and processes described by a master equation. However, in many physical problems it is necessary to consider the case in which the states vary continuously, either because the random variable X is intrinsically continuous (for example, position in space) or because the discretization is so fine that it is more convenient to consider a continuous variation (for example, when the number of individuals in a population becomes very large). In this case, instead of the master equation described in the previous chapter, the evolution of the probability will be governed by a partial differential equation called the *Fokker-Planck equation*.

The derivation of the Fokker-Planck equation starts, as for the master equation, from the Chapman-Kolmogorov equation and proceeds in an almost identical way. For completeness and clarity, we will repeat in the first part of this chapter the necessary steps for its derivation. We will then see some particular cases of great interest and the simplest methods to obtain their solutions. We will also introduce the problem of boundary conditions and thus the problems associated with exit times. The second part of the chapter is dedicated to the complementary approach based on *stochastic differential equations*. We will explicitly see the correspondence with the Fokker-Planck equation and the application to the solution of some simple problems. The chapter concludes with the discussion of a climate model described by a stochastic differential equation.

6.1 Chapman-Kolmogorov Equation for Continuous Processes

Let us consider a continuous variable $X(t)$, which we will assume $X \in \mathbb{R}$, a stochastic function of time $t \in \mathbb{R}$. Let $p(x, t)$ be the probability density defined such that $p(x, t)dx$ is the probability of finding the variable X in the interval $[x, x + dx]$ at time t.

G. Boffetta, A. Vulpiani, *Probability in Physics*, UNITEXT for Physics,
https://doi.org/10.1007/978-3-032-10407-6_6

In general, the process $X(t)$ is defined not only by $p(x, t)$ but also by the possible "correlations," that is, by $W_2(x_1, t_1; x_2, t_2)dx_1dx_2$ (the probability of having the variable in $[x_1, x_1 + dx_1]$ at $t = t_1$ and in $[x_2, x_2 + dx_2]$ at $t = t_2 > t_1$), $W_3(x_1, t_1; x_2, t_2; x_3, t_3)$, and so on. We identify the notation $p(x, t) = W_1(x, t)$, which will be used interchangeably.

An important subset of stochastic processes are those **stationary in time** for which the W_n depend exclusively on the differences of the times (and therefore $p(x, t)$ does not depend on t).

Since they are probabilities, the functions W_n must satisfy

$$W_n \geq 0 \tag{6.1}$$

and it must also hold that

$$W_k(x_1, t_1; \ldots; x_k, t_k) = \int dx_{k+1} \ldots dx_n W_n(x_1, t_1; \ldots; x_n, t_n) \tag{6.2}$$

for every $k < n$, in fact each function W_n must contain the information of the previous W_k. The terms W_n with $n > 1$ are responsible for the memory of the process and thus we can classify stochastic processes (and, as usual, define a Markov process) starting from the relationships that exist among the different W_n.

The simplest case is that of a purely random memoryless process, for which the successive values of x are uncorrelated and thus

$$W_2(x_1, t_1; x_2, t_2) = W_1(x_1, t_1)W_1(x_2, t_2) \tag{6.3}$$

In this case, the process is completely determined by $W_1(x, t) = p(x, t)$ alone. Note that, unlike the discrete-time case, for continuous-time processes it is not easy to find physical examples that satisfy (6.3): in fact, for $t_2 - t_1 \to 0$ in general the states x_1 and x_2 will not be independent, at least if we have some form of continuity in the process.[1] This observation thus leads us to consider the next level of complexity given by Markov processes.

As in the discrete case discussed in Chap. 5, also in the continuous case Markovian processes are defined as having memory only of the previous state. To be more precise, we introduce the *conditional probability* density (or transition probability) $P(x_1, t_1|x_2, t_2)$ defined as the probability density that the random variable X takes a value around x_2 at $t = t_2$ knowing that at time $t_1 < t_2$ it is in state x_1, that is, $P(x_1, t_1|x_2, t_2)dx_2 = P(X(t_2) \in [x_2, x_2 + dx_2]|X(t_1) = x_1)$.[2]

---

[1] An important exception is related to Wiener processes, which we will define later.

[2] We note that, unlike some texts, we use the notation $P(x_1, t_1|x_2, t_2)$ with $t_1 < t_2$.

In the following, we will consider continuous processes, defined by the property that for every $\epsilon > 0$,

$$\lim_{\Delta t \to 0} \frac{1}{\Delta t} \int_{|x-y|>\epsilon} dy P(x, t|y, t + \Delta t) = 0$$

that is, for $\Delta t \to 0$, the probability that y is different from x goes to zero faster than Δt.

From the definition of conditional probability, it must hold that

$$W_2(x_1, t_1; x_2, t_2) = W_1(x_1, t_1) P(x_1, t_1|x_2, t_2)$$

while relations (6.1)–(6.2) imply

$$P(x_1, t_1|x_2, t_2) \geq 0, \tag{6.4}$$

$$\int dx_2 P(x_1, t_1|x_2, t_2) = 1, \tag{6.5}$$

$$p(x_2, t_2) = \int dx_1 p(x_1, t_1) P(x_1, t_1|x_2, t_2). \tag{6.6}$$

which are the analogs of relations (5.1), (5.2) for discrete processes. Similarly, we can define higher-order conditional probabilities, i.e., $P_n(x_1, t_1; x_2, t_2; \ldots |x_n, t_n)$ dx_n (with $n > 2$) is the probability $P(X(t_n) \in [x_n, x_n + dx_n]|X(t_{n-1}) = x_{n-1} \cap \ldots \cap X(t_1) = x_1)$. As in the discrete case, a Markov process is thus defined by the fact that the conditional probabilities of any order $n \geq 2$ are determined only by P (of order 2):

$$P_n(x_1, t_1; x_2, t_2; \ldots |x_n, t_n) = P(x_{n-1}, t_{n-1}|x_n, t_n)$$

and therefore, thanks to (6.6), $P(x_1, t_1|x_2, t_2)$ completely determines the process.

In general, the transition probability $P(x, t|y, t')$ cannot be an arbitrary function of its arguments. In addition to the properties (6.4)–(6.5), it must hold that

$$\lim_{t' \to t} P(x, t|y, t') = \delta(y - x)$$

and P must satisfy the consistency condition given by the **Chapman-Kolmogorov equation**, which in this case is written as

$$P(x, t|y, t') = \int dz P(x, t|z, t_0) P(z, t_0|y, t') \tag{6.7}$$

for every $t_0 \in [t, t']$. The Chapman-Kolmogorov equation encapsulates one of the fundamental aspects of Markov processes: the probability to go from x at time t to y at time t' can be expressed in terms of the probability to go from x to z (at an arbitrary intermediate time) and then from z to y.

6.2 The Fokker-Planck Equation

The Fokker-Planck equation, which governs the evolution of the transition probability $P(x, t|y, t')$ for a very important class of Markov processes, can be obtained in several ways. The simplest is, in analogy with what was done for the master equation, to start from the Chapman-Kolmogorov equation (6.7), which we rewrite in the form

$$P(x_0, t_0|x, t + \Delta t) = \int dz P(x_0, t_0|z, t)P(z, t|x, t + \Delta t) \tag{6.8}$$

where the notation suggests that we will take the limit $\Delta t \to 0$.

Now consider the integral

$$\int dx R(x) \frac{\partial P(x_0, t_0|x, t)}{\partial t}$$

in which $R(x)$ is an arbitrary function that tends to zero sufficiently rapidly for $x \to \pm\infty$. Writing the derivative in terms of the incremental ratio and using (6.8), we have

$$\int dx\ R(x)\ \frac{\partial P(x_0, t_0|x, t)}{\partial t} = \lim_{\Delta t \to 0} \frac{1}{\Delta t} \left[\int dx R(x) \int dz P(z, t|x, t + \Delta t) \times \right.$$
$$\left. \times P(x_0, t_0|z, t) - \int dx R(x) P(x_0, t_0|x, t) \right]. \tag{6.9}$$

In the double integral term in (6.9) we expand in Taylor series $R(x) = R(z) + R'(z)(x - z) + \frac{1}{2}R''(z)(x - z)^2 + \ldots$. The first term of the series, thanks to (6.5), exactly cancels the second integral on the right-hand side of (6.9), and we thus obtain

$$\int dx\ \ R(x) \frac{\partial P(x_0, t_0|x, t)}{\partial t} = \lim_{\Delta t \to 0} \frac{1}{\Delta t} \int dz \int dx \left[R'(z)(x - z) + \right.$$
$$\left. \frac{1}{2} R''(z)(x - z)^2 + \ldots \right] P(z, t|x, t + \Delta t) P(x_0, t_0|z, t).$$

Let us now assume that in a small time interval the variation $(x - z)$ is small and, in particular, that the first two terms of the Taylor series exist and are finite (while the higher order terms vanish in the limit $\Delta t \to 0$). We therefore define

$$\begin{aligned} A(z, t) &= \lim_{\Delta t \to 0} \tfrac{1}{\Delta t} \int dx (x - z) P(z, t|x, t + \Delta t) \\ B(z, t) &= \lim_{\Delta t \to 0} \tfrac{1}{\Delta t} \int dx (x - z)^2 P(z, t|x, t + \Delta t) \end{aligned} \tag{6.10}$$

while we have

$$\lim_{\Delta t \to 0} \frac{1}{\Delta t} \int dx (x - z)^n P(z, t|x, t + \Delta t) = 0 \tag{6.11}$$

for $n \geq 3$ so that we obtain

$$\int dx R(x) \frac{\partial P(x_0, t_0|x, t)}{\partial t} = \int dz P(x_0, t_0|z, t) \left[R'(z) A(z, t) + \frac{1}{2} R''(z) B(z, t) \right].$$

Integrating by parts the second term and relabeling $z \to x$ we have

$$\int dx R(x) \left[\frac{\partial P}{\partial t} + \frac{\partial}{\partial x}(AP) - \frac{1}{2} \frac{\partial^2}{\partial x^2}(BP) \right] = 0. \tag{6.12}$$

Since (6.12) holds for a generic function $R(x)$, the term in square brackets must vanish, and we thus have the general form of the **Fokker-Planck equation** (also called the *forward Kolmogorov equation*)

$$\frac{\partial P(x_0, t_0|x, t)}{\partial t} = -\frac{\partial}{\partial x}(A(x, t) P(x_0, t_0|x, t)) + \frac{1}{2} \frac{\partial^2}{\partial x^2}(B(x, t) P(x_0, t_0|x, t)). \tag{6.13}$$

Note that (6.13), defined over the entire real axis, preserves the condition (6.5).

In a similar way, one can obtain the Fokker-Planck equation *backward* in time, that is, as the evolution with respect to the initial point x_0, which is written as

$$\frac{\partial P(x_0, t_0|x, t)}{\partial t_0} = -A(x_0, t_0) \frac{\partial P(x_0, t_0|x, t)}{\partial x_0} - \frac{1}{2} B(x_0, t_0) \frac{\partial^2 P(x_0, t_0|x, t)}{\partial x_0^2}. \tag{6.14}$$

The Fokker-Planck equation (6.13) also governs the evolution of the probability density $p(x, t)$. In fact, it is sufficient to use (6.6) to obtain

$$\frac{\partial p(x, t)}{\partial t} = -\frac{\partial}{\partial x}(A(x, t) p(x, t)) + \frac{1}{2} \frac{\partial^2}{\partial x^2}(B(x, t) p(x, t)). \tag{6.15}$$

In the multidimensional case (i.e., when $X \in \mathbb{R}^N$) we have, in a completely analogous way,

$$\frac{\partial p(\mathbf{x}, t)}{\partial t} = -\sum_i \frac{\partial}{\partial x_i}(A_i(\mathbf{x}, t)p(\mathbf{x}, t)) + \frac{1}{2}\sum_{i,j} \frac{\partial^2}{\partial x_i \partial x_j}(B_{ij}(\mathbf{x}, t)p(\mathbf{x}, t)) \tag{6.16}$$

where

$$A_i(\mathbf{z}, t) = \lim_{\Delta t \to 0} \frac{1}{\Delta t} \int d\mathbf{x}(x_i - z_i)P(\mathbf{z}, t|\mathbf{x}, t + \Delta t)$$

$$B_{ij}(\mathbf{z}, t) = \lim_{\Delta t \to 0} \frac{1}{\Delta t} \int d\mathbf{x}(x_i - z_i)(x_j - z_j)P(\mathbf{z}, t|\mathbf{x}, t + \Delta t).$$

We observe that in the passage from the Chapman-Kolmogorov equation (6.8) to the Fokker-Planck equation (6.13), we have moved from an integral equation to a partial differential equation. The advantage is evident and is due to the fact that over short times we assume small variations and therefore Eqs. (6.10) and (6.11) hold.

The solution of the Fokker-Planck equation, in the forms (6.13), (6.14), or (6.15), requires knowledge of the initial conditions and boundary conditions. We will address the latter later on. The appropriate initial condition for (6.13) and (6.14) is $P(x_0, t_0|x, t_0) = \delta(x - x_0)$. For (6.15), instead, we will assume a generic initial condition $p(x, t_0) = p_0(x)$.

We conclude this section by noting once again the very close analogy between Eqs. (6.13), (6.15) and the corresponding master equation (5.27), (5.28) derived for discrete-state processes, and by recalling that the former can also be obtained as an appropriate continuous limit of the latter.

6.2.1 *Some Particular Cases*

As an example, let us now consider the particular Fokker-Planck equation obtained directly from two discrete processes considered in the chapter on Markov chains: the random walk and the Ehrenfest model.

In the case of the symmetric random walk, we have seen that the transition matrix is given by

$$W_{ij} = \frac{1}{2}\delta_{i,j-1} + \frac{1}{2}\delta_{i,j+1}. \tag{6.17}$$

Let us now consider the Chapman-Kolmogorov equation in the form (6.8) with Δt the elementary jump time of the process discretized in space with $x = i\Delta x$. $P(x_0|z, \Delta t)$ represents the elementary jump and is therefore defined by the transition matrix (6.17) $P(x_0|z, \Delta t) = W_{ik}$ (with $x_0 = i\Delta x$ and $z = k\Delta x$). We can therefore write, from (6.8)

$$P(x_0|x, t + \Delta t) = \frac{1}{2} \left[P(x_0|x + \Delta x, t) + P(x_0|x - \Delta x, t) \right]. \tag{6.18}$$

We note that the interpretation of (6.18) is transparent: the probability of arriving at point x at time $t + \Delta t$ is given by the probability of being at an adjacent site times the probability of jumping (which is $1/2$). Now, subtracting $P(x_0|x, t)$ from both sides and dividing by Δt, we have

$$\frac{P(x_0|x, t + \Delta t) - P(x_0|x, t)}{\Delta t} = D \frac{P(x_0|x + \Delta x, t) + P(x_0|x - \Delta x, t) - 2P(x_0|x, t)}{\Delta x^2} \tag{6.19}$$

where we have introduced the **diffusion coefficient** $D \equiv \frac{\Delta x^2}{2\Delta t}$.

In order for the expression (6.19) to have a nontrivial limit as $\Delta t \to 0$, we must simultaneously take the limit $\Delta t \to 0$ and $\Delta x \to 0$ with D constant, thus obtaining the well-known continuous limit of the random walk given by the **diffusion equation** (or heat equation)

$$\frac{\partial P(x_0|x, t)}{\partial t} = D \frac{\partial^2 P(x_0|x, t)}{\partial x^2} \tag{6.20}$$

which is nothing but the Fokker-Planck equation (6.13) in the particular case where $A(x) = 0$ and $B(x) = 2D$. This example thus allows us to identify the coefficient B in (6.13) as a diffusion coefficient. As for the coefficient A, we must consider a more general case in which the probabilities of jumping to the right and to the left are not equal.

Let us therefore take an asymmetric random walk in which the transition matrix is given by

$$W_{ij} = \left(\frac{1}{2} + \varepsilon \right) \delta_{i,j-1} + \left(\frac{1}{2} - \varepsilon \right) \delta_{i,j+1}.$$

Proceeding as in the previous case, we obtain from (6.8)

$$\frac{P(x_0|x, t + \Delta t) - P(x_0|x, t)}{\Delta t} = -\frac{\varepsilon \Delta x}{\Delta t} \frac{P(x_0|x + \Delta x, t) - P(x_0|x - \Delta x, t)}{\Delta x} + \frac{\Delta x^2}{2\Delta t} \frac{P(x_0|x + \Delta x, t) + P(x_0|x - \Delta x, t) - 2P(x_0|x, t)}{\Delta x^2}. \tag{6.21}$$

We now want to take the continuous limit so that both terms on the right in (6.21) remain finite. To achieve this, we must require that ε is $O(\Delta x)$, that is, we define

$\varepsilon = \frac{\beta}{2D}\Delta x$ in order to obtain

$$\frac{\partial P(x_0|x,t)}{\partial t} = -\beta \frac{\partial P(x_0|x,t)}{\partial x} + D \frac{\partial^2 P(x_0|x,t)}{\partial x^2} \tag{6.22}$$

where β, and therefore in general $A(x,t)$, defines the **drift coefficient**. We note that both Eqs. (6.20) and (6.22), being particular cases of (6.13), also govern the evolution of the density $p(x,t)$.

Let us finally consider the more complex example of the Ehrenfest model. Referring to the notation of Sect. 5.2.3, we consider the continuous variable x centered on the equilibrium value $N/2$, that is, we define $x = (k - N/2)\Delta x$ and $L = N\Delta x$ which represents the length of the domain. Again, starting from the Chapman-Kolmogorov equation (6.8) with $P(x_0|z, \Delta t)$ given by the transition matrix (5.16) we can write

$$P(x_0|x, t + \Delta t) = \left(\frac{1}{2} + \frac{x + \Delta x}{L}\right) P(x_0|x + \Delta x, t) + \\ + \left(\frac{1}{2} + \frac{x - \Delta x}{L}\right) P(x_0|x - \Delta x, t)$$

Again, subtracting $P(x_0|x,t)$ we obtain the Fokker-Planck equation in the form known as the **Ornstein-Uhlenbeck process**

$$\frac{\partial P(x_0|x,t)}{\partial t} = k\frac{\partial}{\partial x}(xP(x_0|x,t)) + D \frac{\partial^2 P(x_0|x,t)}{\partial x^2} \tag{6.23}$$

where the drift term (with $k \equiv \frac{\Delta x}{L\Delta t}$) represents the effect of an attractive force toward the center $x = 0$.

6.2.2 *Solutions by the Fourier Transform Method*

Returning to the general case of the Fokker-Planck equation in the form (6.15), the fundamental problem of determining the probability distribution $p(x,t)$ thus reduces to finding the solution of the partial differential equation (6.13) with initial condition given by the Dirac delta function $p(x, 0) = \delta(x - x_0)$. In the case where the coefficients A and B are constant, the solution can be easily obtained by means of Fourier transforms.

Recall that the Fourier transform (FT) of a function $p(x)$ (which we assume to be L^2) is defined as

$$\hat{p}(k) = \frac{1}{\sqrt{2\pi}} \int_{-\infty}^{+\infty} e^{ikx} p(x) dx$$

while the inverse transform is given by

$$p(x) = \frac{1}{\sqrt{2\pi}} \int_{-\infty}^{+\infty} e^{-ikx} \hat{p}(k) dk.$$

We now apply the FT to (6.20) and obtain

$$\frac{\partial \hat{p}(k,t)}{\partial t} = -Dk^2 \hat{p}(k,t)$$

which therefore has the solution $\hat{p}(k,t) = \hat{p}(k,0)e^{-Dk^2 t}$. The initial condition $p(x,0) = \delta(x - x_0)$ has FT given by $\hat{p}(k,0) = (1/\sqrt{2\pi})e^{ikx_0}$. Substituting and taking the inverse transform we obtain

$$p(x,t) = \frac{1}{2\pi} \int_{-\infty}^{+\infty} e^{-ik(x-x_0)-Dk^2 t}$$

which, when integrated, leads to the Gaussian solution[3]

$$p(x,t) = \frac{1}{\sqrt{4\pi Dt}} \exp\left[-\frac{(x-x_0)^2}{4Dt}\right]$$

with mean value x_0 and variance $2Dt$, identical to that found by taking the continuous limit of the random walk in Chap. 4.

In the case of the equation with drift (6.22), using a similar method we obtain

$$p(x,t) = \frac{1}{(4\pi Dt)^{1/2}} \exp\left[-\frac{(x-(x_0+\beta t))^2}{4Dt}\right]$$

that is, a Gaussian with mean value driven by the drift $\langle x(t) \rangle = x_0 + \beta t$ and variance that grows with the diffusion coefficient as $\sigma^2(t) = \langle x^2 \rangle - \langle x \rangle^2 = 2Dt$.

In the case of the Ornstein-Uhlenbeck equation (6.23) the FT method is not applicable since the drift coefficient is not constant. The solution, however, can be obtained in an elementary way as

$$p(x,t) = \frac{1}{(2\pi\sigma^2)^{1/2}} \exp\left[-\frac{(x-\bar{x})^2}{2\sigma^2}\right] \tag{6.24}$$

where $\bar{x} = \langle x \rangle = x_0 \exp(-kt)$ represents the mean value mean and $\sigma^2 = \frac{D}{k}[1 - \exp(-2kt)]$. The distribution is still Gaussian, but the mean shifts towards zero and the variance tends to a constant value. In the limit $t \to \infty$ the

[3] Recall the useful formula for the Gaussian integral $\int_{-\infty}^{+\infty} e^{-ax^2+bx} dx = \sqrt{\frac{\pi}{a}} e^{b^2/(4a)}$.

distribution (6.24) obviously reproduces the continuous limit of the stationary discrete distribution (5.18).

Let us observe that the Fokker-Planck equation can also be obtained starting from the continuous limit of the master equation. Let us consider the birth-death problem discussed in Sect. 5.4.1. The master equation for the probability distribution $p_n(t)$ is given by (5.29)

$$\frac{dp_n(t)}{dt} = (n+1)p_{n+1}(t) - np_n(t). \tag{6.25}$$

In the limit of large populations, where the variation of the variable n can be considered continuous, (6.25) obviously becomes

$$\frac{\partial p(n,t)}{\partial t} = \frac{\partial}{\partial n}(np(n,t))$$

which is nothing but an Ornstein-Uhlenbeck process with drift $k = 1$ and without diffusion. From what we have just seen, the evolution of the mean value starting from the initial condition $n_0 = N$ will be therefore $\langle n \rangle = Ne^{-t}$ as found in Chap. 5.

6.3 The Fokker-Planck Equation with Barriers

Let us rewrite the FP equation (6.15), which governs the evolution of the probability density, in the form

$$\frac{\partial p(x,t)}{\partial t} + \frac{\partial J(x,t)}{\partial x} = 0$$

where we have introduced the **probability current**

$$J(x,t) = A(x,t)p(x,t) - \frac{1}{2}\frac{\partial}{\partial x}(B(x,t)p(x,t)).$$

In this way, the FP equation takes the form of a local (conservation) equation (of probability). Let us consider a region $I = [a,b] \subset \mathbb{R}$ and define the probability of being in the interval I

$$P_I(t) = \int_I p(x,t)dx, \tag{6.26}$$

we will obviously have

$$\frac{\partial P_I(t)}{\partial t} = -J(b,t) + J(a,t)$$

that is, the probability P_I increases if $J(a,t) > J(b,t)$ and thus J represents the flux in the direction of positive x. We can now define different boundary conditions on the boundary of I, $\partial I = \{a, b\}$.

Reflecting Barriers In this case, we require that the probability of leaving (and entering) I is zero, therefore the current at the boundary must be zero

$$J(x,t) = 0 \text{ for } x \in \partial I$$

Note that in this case the total probability in I, $P_I(t)$, is conserved.

Absorbing Barriers In the case of absorbing barriers, we want that when the particle reaches the boundary of I it is removed from the system. Therefore, the probability of finding the particle in ∂I must vanish and thus

$$p(x,t) = 0 \text{ for } x \in \partial I$$

Obviously, in this case the probability (6.26) is not conserved.

Periodic Boundary Conditions Another interesting case, and one widely used in numerical simulations, is the periodic case: if the particle exits from one side of the interval, it re-enters from the other (and thus I is defined on a circle). In this case, we will have

$$p(b,t) = p(a,t) \text{ and } J(b,t) = J(a,t).$$

Often, periodic boundary conditions are imposed together with the periodicity of the coefficients $A(b,t) = A(a,t)$ and $B(b,t) = B(a,t)$.

Let us also mention that the case "without barriers" corresponds to the case in which $I = \mathbb{R}$ and for which we must assume that the probability decays sufficiently rapidly for $x \to \pm\infty$ in order to guarantee normalization and the existence of the moments.

6.3.1 Stationary Solutions of the Fokker-Planck Equation

Let us now consider the problem of the time-independent solution of the Fokker-Planck equation, that is, for which

$$p(x,t) = \pi(x)$$

where the stationary density $\pi(x)$ is a solution of the stationary FP equation

$$-\frac{\partial}{\partial x}(A(x)\pi(x)) + \frac{1}{2}\frac{\partial^2}{\partial x^2}(B(x)\pi(x)) = 0 \tag{6.27}$$

with the appropriate boundary conditions. In (6.27) we have assumed that the process is *homogeneous* and therefore the coefficients A and B do not depend on time.

A relevant problem, which we have already discussed in the case of Markov chains, is to find the conditions under which a generic initial probability density $p_0(x)$ evolves towards the stationary density $\pi(x)$. A general result guarantees convergence to $\pi(x)$ when the diffusion coefficient $B(x)$ is everywhere positive. In Chap. 4 this property was shown explicitly in the case of the discrete-time model for Brownian motion.

Let us now look at some simple examples of stationary solutions of the Fokker-Planck equation. We start from the simplest case of a purely diffusive process (6.20), that is, with $A(x) = 0$ and $B(x) = 2D$ defined in the interval $I = [-a, a]$ with reflecting boundary conditions at $x = \pm a$. The solution of (6.27) in this case is obviously

$$\pi(x) = \frac{1}{2a}$$

that is, the probability of finding the particle in the interval $[-a, a]$ is uniform. Note that, obviously, the flux is everywhere zero: $J(x) = -D\frac{d\pi(x)}{dx} = 0$.

In the case of absorbing boundary conditions, the stationary solution is obviously $\pi(x) = 0$: in fact, at long times all the particles will have ended up on one of the barriers. To have non-trivial situations, we can imagine replacing the particles absorbed by the walls.

Let us therefore consider the case in which at each unit of time we release a particle at the point $x_0 \in [a, b]$. The solution of the stationary equation will be of the form $\pi(x) = a_1 + a_2 x$ and we want that at $x = \pm a$ it is $\pi(x) = 0$. We will then have

$$\pi(x) = \begin{cases} C_1(a - x) \; x \geq x_0 \\ C_2(a + x) \; x \leq x_0 \end{cases} \tag{6.28}$$

Continuity at x_0, i.e., $\pi(x_0^-) = \pi(x_0^+)$ leads to the condition $C_2 = C_1 \frac{a-x_0}{a+x_0}$, while the value of the coefficient C_1 is determined by the total number of particles in the interval I, and therefore by the release frequency.[4]

Let us now consider the more complicated case of the (homogeneous) Ornstein-Uhlenbeck process (6.23), that is, with $A(x) = -kx$ and $B(x) = 2D$ and again defined in the interval $I = [-a, a]$. Assuming reflecting boundary conditions at $x = \pm a$, the solution of (6.27) is (as can be directly verified)

$$\pi(x) = Ce^{\frac{k}{2D}(a^2 - x^2)}$$

[4] Note that while $\pi(x)$ is continuous at x_0, this is not true for its derivative and therefore for the flux $J(x)$. In fact, the point x_0 is a "source" of particles.

where C is a constant determined by the normalization condition $\int_{-a}^{a} \pi(x)dx = 1$. Note that the flux $J(x) = -kx\pi - D\frac{\partial \pi}{\partial x}$ is zero throughout I in this case as well. In the case of absorbing boundary conditions, the stationary solution is the trivial one $\pi(x) = 0$.

A physically interesting and easily treatable case is that of a process with conditions at infinity $\pi(x) \to 0$ for $x \to \pm\infty$ with constant diffusion coefficient, $B(x) = 2D$. Writing $A(x)$ in terms of a "potential"

$$A(x) = -\frac{dV(x)}{dx} \tag{6.29}$$

which we will assume satisfies $\lim_{x\to\pm\infty} V(x) = \infty$, Eq. (6.27) is easily solvable with solution

$$\pi(x) = Ce^{-V(x)/D}$$

where the constant C is fixed by normalization. This solution can be generalized to the multidimensional case when it is possible to write $A_i(x) = -\frac{\partial}{\partial x_i} V(x)$, in which case we again have

$$\pi(\mathbf{x}) = Ce^{-V(\mathbf{x})/D}.$$

Obviously, in one dimension it is always possible to write $A(x)$ in the form (6.29).

6.3.2 Exit Times for Homogeneous Processes

An interesting problem for many applications is the calculation of **exit times** or first passage times. We have already discussed this issue in the case of Markov chains for the *gambler's ruin* problem (see Sect. 5.1.2).

Let us therefore consider the simple case of a purely diffusive process (6.20) (continuous limit of the gambler's problem) with absorbing barriers at $x = \pm a$ and with initial condition $X(0) = x_0$. The stationary solution is given by (6.28) with $C_2 = C_1\frac{a-x_0}{a+x_0}$. The particle flux through the two barriers is given by $J(x) = -D\frac{\partial \pi(x)}{\partial x}$, that is,

$$J(-a) = -DC_1\frac{a-x_0}{a+x_0}\ , \ J(a) = DC_1.$$

Therefore, the probability that the particle, starting from x_0, diffuses through the barrier at $-a$ is given by

$$P_{x_0}(-a) = \frac{-J(-a)}{-J(-a) + J(a)} = \frac{1}{2} - \frac{x_0}{2a} \tag{6.30}$$

which is nothing but the continuous version of the result (5.10) obtained for Markov chains. Note that in the denominator of (6.30) we have the total particle flux $J_{tot} = -J(-a) + J(a) = \frac{2aC_1 D}{a+x_0}$, which represents the number of particles leaving the domain I per unit time. The total number of particles present in the domain will be given by

$$N = \int_{-a}^{a} \pi(x)dx = C_1 a(a - x_0)$$

and therefore the *mean exit time* starting from x_0 will be

$$T(x_0) = \frac{N}{J_{tot}} = \frac{a^2 - x_0^2}{2D}. \tag{6.31}$$

Let us now consider the more general case, which is of great practical interest, of a diffusive process in a potential, for which the Fokker-Planck equation (6.15) can be written as

$$\frac{\partial p(x,t)}{\partial t} = +\frac{\partial}{\partial x}\left(\frac{dV(x)}{dx}p(x,t)\right) + D\frac{\partial^2}{\partial x^2}p(x,t)$$

with $A(x) = -\frac{dV(x)}{dt}$. Given an interval $I = [a, b]$, let us define $G(x, t)$ as the probability of still being in the interval I after a time t, assuming we started from x at the initial time $t_0 = 0$ with $a \leq x \leq b$

$$G(x,t) = \int_a^b P(x, t_0|y, t)dy.$$

Let $T(x)$ be the exit time from I; we will obviously have that

$$P(T(x) > t) = G(x,t) = \int_a^b P(x, t_0|y, t)dy.$$

The equation governing the evolution of $G(x, t)$, and thus of $T(x)$, with x being the starting point, will be closely related to the backward Fokker-Planck equation (6.14), which we rewrite as

$$\frac{\partial P(x, t_0|y, t)}{\partial t_0} = -A(x)\frac{\partial P(x, t_0|y, t)}{\partial x} - D\frac{\partial^2 P(x, t_0|y, t)}{\partial x^2}. \tag{6.32}$$

But since the process is stationary, we have $\partial_{t_0} P(x, t_0|y, t) = -\partial_t P(x, t_0|y, t)$, which, substituted into (6.32) and then integrating over x from a to b, becomes

$$\frac{\partial G(x,t)}{\partial t} = A(x)\frac{\partial G(x,t)}{\partial x} + D\frac{\partial^2 G(x,t)}{\partial x^2}. \tag{6.33}$$

By definition, $G(x, t)$ is the probability that the exit time from I is greater than t, so denoting by $\Pi(T)$ the probability density of exit times, we have $G(x, t) = \int_t^\infty \Pi(T)dT$, that is, $\Pi(t) = -\frac{\partial G(x,t)}{\partial t}$. The *mean first exit time* will be

$$T(x) = \langle T \rangle = \int_0^\infty t\Pi(t)dt = -\int_0^\infty t\partial_t G(x, t)dt = \int_0^\infty G(x, t)dt \tag{6.34}$$

where in the last step we have integrated by parts. Now, noting that $G(x, t_0) = 1$ (if $x \in I$) while $G(x, \infty) = 0$, integrating (6.33) over t from 0 to ∞ and using (6.34), we obtain the equation for the **mean first exit time** $T(x)$

$$A(x)\frac{\partial T(x)}{\partial x} + D\frac{\partial^2 T(x)}{\partial x^2} = -1. \tag{6.35}$$

As a first example of application of (6.35), let us derive again the result (6.31) for the mean first exit time of a diffusive process. Let us therefore consider absorbing barriers at the endpoints $x = \pm a$, for which the boundary conditions for (6.35) will be $T(-a) = T(a) = 0$ (obviously: if I start at the boundary, I exit immediately). Since $A(x) = 0$, Eq. (6.35) becomes $D\partial_x^2 T(x) = -1$, which has the solution

$$T(x) = \frac{1}{D}(c_0 + c_1 x - x^2/2)$$

The coefficients are determined by the boundary conditions as $c_0 = a^2/2$ and $c_1 = 0$, and thus we recover (6.31).

The Kramers Approximation

An important application of Eq. (6.35) is the calculation of the exit time over a potential barrier, a simple mathematical model for many processes, such as the Arrhenius law for chemical reactions.

Let us consider the case in which $V(x)$ is a potential with two minima (not necessarily symmetric) separated by a potential barrier, as schematically illustrated in Fig. 6.1. We assume that the potential increases indefinitely for $x \to \pm\infty$ in order to guarantee the existence of a stationary distribution (which will obviously be peaked around the two minima). As a physical model, consider a Brownian particle confined in the potential and initially located near one of the two minima (for example, in the left well). We want to determine the mean time it takes for the Brownian particle to cross the potential and end up in the right well. Clearly, this time depends on the height of the potential barrier, but, as we will see, it does not depend much on the details of the shape of the potential.

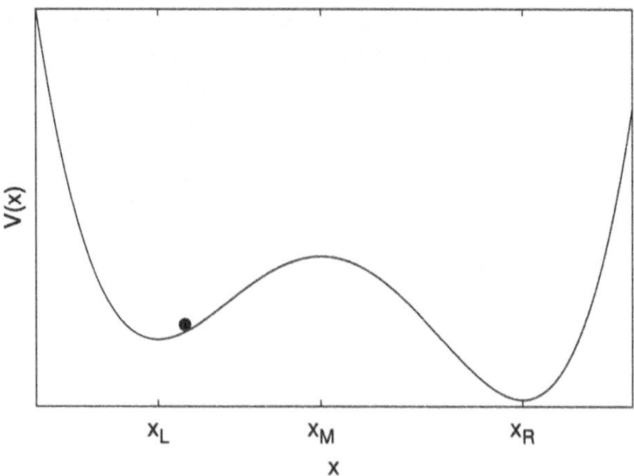

Fig. 6.1 Brownian particle in a double-well potential

Since at time $t_0 = 0$ the particle is at x near the left minimum x_L, calculating the mean time it takes to reach x_R is a first-passage problem with an absorbing barrier at $b = x_R$ and a reflecting barrier at $a = -\infty$. We therefore seek the solution of (6.35) with $T'(a) = T(b) = 0$ ($T'(x) = dT/dx$). Introducing the auxiliary function $\psi(x)$ defined by $A(x) \equiv D\psi'(x)/\psi(x)$, (6.35) can be rewritten as

$$\frac{d}{dx}(\psi(x)T'(x)) = -\frac{1}{D}\psi(x)$$

which, integrated between $-\infty$ (where $\psi = 0$) and x, yields

$$T'(x) = -\frac{1}{D}\frac{1}{\psi(x)}\int_{-\infty}^{x}\psi(z)dz$$

Integrating further (and using $T(x_R) = 0$), we finally obtain

$$T(x) = \frac{1}{D}\int_{x}^{x_R}\frac{dy}{\psi(y)}\int_{-\infty}^{y}\psi(z)dz. \tag{6.36}$$

In our case, where $A(x) = -V'(x)$, the solution (6.36) can be written as

$$T(x) = \frac{1}{D}\int_{x}^{x_R} dy\, e^{\frac{V(y)}{D}}\int_{-\infty}^{y} dz\, e^{-\frac{V(z)}{D}}.$$

Let us now observe that the first integrand, $e^{V(y)/D}$, is dominated by values of y near the maximum, $y \simeq x_M$, where the second integral has a weak dependence on y (because for $z \simeq x_M$ the integrand $e^{-V(z)/D}$ is very small), and therefore we can replace y with x_M in the second integral and obtain

$$T(x) \simeq \frac{1}{D} \int_{-\infty}^{x_M} dz\, e^{-\frac{V(z)}{D}} \int_x^{x_R} dy\, e^{\frac{V(y)}{D}}.$$

By construction, the potential has a maximum at x_M and a minimum at x_L, which we will assume to be quadratic, and therefore we can assume that for $x \simeq x_M$,

$$V(x) \simeq V(x_M) - \frac{(x - x_M)^2}{2\delta^2}$$

while for $x \simeq x_L$

$$V(x) \simeq V(x_L) + \frac{(x - x_L)^2}{2\alpha^2}.$$

At this point we have

$$\int_{-\infty}^{x_M} dz\, e^{-\frac{V(z)}{D}} \simeq e^{-\frac{V(x_L)}{D}} \int_{-\infty}^{x_M} dz\, e^{-\frac{(z-x_L)^2}{2D\alpha^2}} \simeq \alpha\sqrt{2\pi D} e^{-\frac{V(x_L)}{D}}$$

where in the last step we have replaced x_M with ∞ in the upper limit of the integral since it is dominated by $z \simeq x_M$. Similarly,

$$\int_x^{x_R} dy\, e^{\frac{V(y)}{D}} \simeq e^{\frac{V(x_M)}{D}} \int_x^{x_R} dy\, e^{-\frac{(y-x_M)^2}{2D\delta^2}} \simeq \delta\sqrt{2\pi D} e^{\frac{V(x_M)}{D}}$$

where again we have used the fact that the integral is dominated by $y \simeq x_M$. In conclusion, we obtain the **Kramers formula** for the mean exit time from the potential well (starting from a point $x < x_M$ to a point $x' > x_M$)

$$T(x) = 2\pi\alpha\delta\, e^{\frac{\Delta V}{D}}. \tag{6.37}$$

We observe that, as anticipated, the exit time depends strongly on the height of the potential $\Delta V = V(x_M) - V(x_L)$ and only weakly on the details of its shape, appearing only in the constants in front of the exponential term in (6.37). Obviously, for $D \to 0$ the particle will be confined in the initial well for an exponentially long time.

In chemical reaction theory, (6.37) is known as the **Arrhenius formula**. In this case, the variable x represents the chemical species: $x = x_L$ is species A while $x = x_R$ is species B in the reaction $A \to B$, separated by a potential barrier ΔE. The diffusion coefficient in this case is $D = kT$ and therefore the reaction rate (inverse of the exit time) will be proportional to $e^{-\Delta E/(kT)}$, which is the Arrhenius law.

6.4 Stochastic Differential Equations

The Fokker-Planck equation (6.13) is a deterministic equation that describes the evolution of the probability density. We now want to consider the time evolution of a single realization of the stochastic process $X(t)$, and for this we need a **stochastic differential equation** (SDE). We have already encountered stochastic differential equations in the Langevin model for Brownian motion (4.1) in Chap. 4, in the form

$$m\frac{dv}{dt} = -6\pi a\mu v + \xi. \tag{6.38}$$

The fundamental characteristic of an SDE is the simultaneous presence of a deterministic term (the Stokes force, in (6.38)) and a stochastic term (the force due to the molecules) $\xi(t)$. We assume that this term has zero mean $\langle\xi(t)\rangle = 0$ and zero correlation time, that is,[5]

$$\langle\xi(t)\xi(t')\rangle = const.\delta(t - t'). \tag{6.39}$$

To give a mathematically precise interpretation to $\xi(t)$, let us consider the integral quantity

$$W(t) = \int_0^t dt'\xi(t')$$

for which we obviously have $\langle W(t)\rangle = 0$, while from (6.39) we obtain $\langle W^2(t)\rangle = t$. The stochastic process $W(t)$ is called the **Wiener process** (or random walk in the continuous limit) and is characterized by the following properties (with $W(0) = 0$):

1. $W(t)$ is continuous;
2. for any $t_1 < t_2 < t_3$, the increments $W(t_3) - W(t_2)$ and $W(t_2) - W(t_1)$ are independent;
3. for any $t_1 < t_2$, the increment $W(t_2) - W(t_1)$ has a Gaussian distribution with zero mean and variance $t_2 - t_1$.

We note that $W(t)$ is continuous but not differentiable (in fact, the "derivative" $\xi(t)$ is not a function in the usual sense). Indeed, let us compute the probability that the modulus of the incremental ratio $\frac{W(t+\Delta t)-W(t)}{\Delta t}$ is greater than a given $\delta > 0$, that is, the probability that $|W(t + \Delta t) - W(t)|$ is greater than $\delta\Delta t$. By property (3) of the

[5] From a physical point of view, assuming a zero correlation time is justified if there is a separation of scales, in which the dynamics of $\xi(t)$ occurs on timescales much faster than the evolution of the variable $X(t)$.

Wiener process we have

$$\lim_{\Delta t \to 0} P\left(\frac{|W(t+\Delta t) - W(t)|}{\Delta t} > \delta\right) = \lim_{\Delta t \to 0} \frac{2}{\sqrt{2\pi \Delta t}} \int_{\delta \Delta t}^{\infty} e^{-\frac{y^2}{2\Delta t}} dy = 1$$

and therefore, since this holds for arbitrary δ, the derivative of $W(t)$ is almost everywhere undefined.

At this point, we can write the prototype of a **stochastic differential equation** of diffusive type in the form

$$dX(t) = a(X(t), t)dt + b(X(t), t)dW(t) \tag{6.40}$$

where we do not divide by dt to remind ourselves that dW is not a differential. In practice, as seen before, we can think of dW as $Z(t)\sqrt{dt}$ where $Z(t)$ is a time-independent process with zero mean and unit variance. We will see later that (6.40) is equivalent to the Fokker-Planck equation (6.13). On a heuristic level, we can observe that

$$\begin{aligned}
\langle dX(t)\rangle &= a(X(t), t)dt \\
\langle (dX(t))^2\rangle &= a^2(X(t), t)dt^2 + 2a(X(t), t)b(X(t), t)\langle dW\rangle dt + \\
&\quad b^2(X(t), t)\langle dW^2\rangle \simeq b^2(X(t), t)dt
\end{aligned}$$

up to higher order terms in dt, and therefore we have that the Fokker-Planck equation for the probability density of $X(t)$ is obtained by setting $A(x, t) = a(x, t)$ and $B(x, t) = b^2(x, t)$.

6.4.1 *Ito's Formula*

The evolution of the stochastic variable $X(t)$ will be given by the integration of the SDE (6.40), which formally leads to

$$X(t) = X(0) + \int_0^t a(X(s), s)ds + \int_0^t b(X(s), s)dW(s).$$

The calculation of the last integral is a delicate point due to the presence of the dW term. Recall that (deterministic) integrals are usually introduced as the limit of a discrete sum. The discretization can be done in different ways, for example according to the rectangular (Euler) rule, or the trapezoidal rule. In the case of integrals of "ordinary" functions (i.e., non-random), the chosen procedure is irrelevant because in the continuous limit all discretizations converge to the value that defines the integral. In the case of stochastic integrals, the issue is more subtle since the value assumed in the continuous limit can depend on the adopted discretization. As an example, let us consider the calculation of the simple stochastic integral $I = \int_0^t W(s)dW(s)$. The discretization according to the "forward" Euler rule is written (using the notation $\Delta W(t_i) = W(t_{i+1}) - W(t_i)$)

$$I = \sum_{i=0}^{N-1} W_i \Delta W(t_i) \tag{6.41}$$

whose mean value, by property (2) of the Wiener process, is $\langle I \rangle = 0$. If instead we consider the "backward" Euler rule, we obtain

$$I = \sum_{i=0}^{N-1} W(t_{i+1}) \Delta W(t_i). \tag{6.42}$$

Let us now compute the mean

$$\langle I \rangle = \sum_{i=0}^{N-1} \langle (W(t_i) + \Delta W(t_i)) \Delta W(t_i) \rangle = \sum_{i=0}^{N-1} \langle (\Delta W(t_i))^2 \rangle = t$$

where the last step uses property (3). This simple example shows how extremely important the choice of discretization used for the calculation of stochastic integrals is. The discretization (6.41) is known as the *Ito* rule, while (6.42) is called the *Stratonovich* rule.

For convenience, we summarize below some results related to the calculation of the most common Ito stochastic integrals:

$$\langle \int_0^t f(s)dW(s) \rangle = 0 \tag{6.43}$$

$$\langle \int_0^t W(s)dW(s) \rangle = 0$$

$$\langle \int_0^t f(s)dW(s) \int_0^t f(s')dW(s') \rangle = \int_0^t f^2(s)ds$$

where $f(t)$ is any function, and for the last result we have made the formal substitution $\langle dW(s)dW(s') \rangle = \delta(s - s')dsds'$.

A fundamental result related to stochastic differential equations is the so-called **Ito formula**, which allows a change of variable between $X(t)$, which obeys the diffusive process (6.40), and the variable $Y(t) = f(X(t), t)$ (where $f(x, t)$ is a continuous and differentiable function, at least twice in x):

$$dY(t) = \left(\frac{\partial f}{\partial t} + a \frac{\partial f}{\partial x} + \frac{1}{2} b^2 \frac{\partial^2 f}{\partial x^2} \right) dt + b \frac{\partial f}{\partial x} dW(t). \tag{6.44}$$

The heuristic derivation of (6.44) is not difficult: it is enough to expand $f(x,t)$ in a Taylor series in x and t

$$df = \frac{\partial f}{\partial t}dt + \frac{\partial f}{\partial x}dX + \frac{1}{2}\frac{\partial^2 f}{\partial x^2}dX^2 \tag{6.45}$$

and substitute, in accordance with (6.40), $dX = adt + bdW$ and keep the terms of order dt and setting $dW^2 = dt$ (which is a sort of "rule of thumb" justified by the fact that $\langle dW^2\rangle = dt$). Ito's formula is nothing more than the generalization of the formula for the derivative of composite functions (to which it reduces in the case where $b = 0$), and it tells us that the derived stochastic variable $Y(t)$ also obeys an SDE of the type (6.40). Its use allows us to reduce many problems to known stochastic processes and thus find their solution.

6.4.2 From the SDE to the Fokker-Planck Equation

As a first application of Ito's formula, let us derive the connection between the SDE (6.40) and the Fokker-Planck equation for the evolution of probability. Given an arbitrary function $f(x)$ (not explicitly dependent on t), Eq. (6.44) can be written as

$$d\,f = \left(a\frac{\partial f}{\partial x} + \frac{1}{2}b^2\frac{\partial^2 f}{\partial x^2}\right)dt + b\frac{\partial f}{\partial x}dW(t)$$

Taking the average (recalling that $\langle dW\rangle = 0$), we obtain

$$\frac{d}{dt}\langle f(x)\rangle = \langle a\frac{\partial f}{\partial x} + \frac{1}{2}b^2\frac{\partial^2 f}{\partial x^2}\rangle. \tag{6.46}$$

Introducing the probability density $p(x,t)$ in the definition of the average and integrating by parts the right-hand side of (6.46), we obtain

$$\int dxf(x)\frac{\partial p(x,t)}{\partial t} = \int dxf(x)\left[-\frac{\partial}{\partial x}(a(x,t)p(x,t)) + \right.$$
$$\left. + \frac{1}{2}\frac{\partial^2}{\partial x^2}(b^2(x,t)p(x,t))\right]$$

which, since it must hold for arbitrary $f(x)$, implies (6.13) with $A = a$ and $B = b^2$. In general, therefore, the SDE of the type (6.40) describes a diffusive process with drift coefficient a and diffusion coefficient b^2.

Note that there is no real conflict between the Ito and Stratonovich methods. If the stochastic differential equation is interpreted in the Stratonovich sense, the drift coefficient a^S will be given by the coefficients a^I and b of the Ito equation by the relation $a^S = a^I - \frac{1}{2}b\partial_x b$.

We can also say that, given a Fokker-Planck equation, one can unambiguously write two stochastic differential equations following either the Ito or Stratonovich interpretation. Note, however, that if the diffusion coefficient b is constant, the distinction between the two interpretations disappears.

6.4.3 The Ornstein-Uhlenbeck Process

As a further example of the application of Ito's formula, let us consider the Ornstein-Uhlenbeck process discussed in Sect. 6.2.1. The stochastic differential equation governing it is written as

$$dX(t) = -kX(t)dt + \sqrt{2D}dW(t).$$

With the change of variables $Y(t) = f(X) = X(t)e^{kt}$, from (6.44) we obtain

$$dY(t) = \sqrt{2D}e^{kt}dW(t)$$

which, when integrated, leads to the formal solution

$$Y(t) = Y(0) + \sqrt{2D}\int_0^t e^{ks}dW(s)$$

or, in the original variables,

$$X(t) = X(0)e^{-kt} + \sqrt{2D}\int_0^t e^{-k(t-s)}dW(s).$$

Let us now consider the mean quantities. Using (6.43) we have for the mean value

$$\langle X(t)\rangle = X(0)e^{-kt}$$

while for the variance

$$\sigma^2(t) = \langle (X(t) - \langle X(t)\rangle)^2\rangle = 2D\langle \int_0^t e^{-k(t-s)}dW(s)\int_0^t e^{-k(t-s')}dW(s')\rangle =$$

$$= 2D\int_0^t e^{-2k(t-s)}ds = \frac{D}{k}\left(1 - e^{-2kt}\right)$$

in agreement with what was found with (6.24). The Ornstein-Uhlenbeck process is important because it represents the simplest example of the motion of a particle in a potential subject to thermal noise. In the case of Ornstein-Uhlenbeck, the potential is parabolic ($V(x) = \frac{1}{2}kx^2$), while in the general case we will have the SDE $dX(t) = -\frac{\partial V}{\partial x}dt + \sqrt{2D}dW(t)$.

6.4.4 Geometric Brownian Motion

Let us now discuss a less elementary example of the application of Ito's formula of Ito to the *geometric Brownian motion*, defined by the stochastic process for the variable $S(t)$

$$dS(t) = S(t)(\mu dt + \sigma dW(t)) \qquad (6.47)$$

(where μ and σ are two constants). Geometric Brownian motion is a fundamental ingredient in the applications of stochastic processes to financial problems, as it is based on the (simple but in some ways reasonable) assumption that the *relative* variation dS/S (of a price, in this specific case) obeys a diffusive process. Applying Ito's formula to the function $Y(t) = f(S) = \log(S(t))$ we have

$$dY = \frac{\partial f}{\partial S}dS + \frac{1}{2}\frac{\partial^2 f}{\partial S^2}dS^2.$$

Now substituting (6.47) and using $dW^2 = dt$ we obtain

$$dY(t) = (\mu - \frac{\sigma^2}{2})dt + \sigma dW(t)$$

which is again a diffusive process of the type described by (6.40). Therefore, in geometric Brownian motion, the logarithm of the variable S follows a diffusive process (that is, $S(t)$ follows a *log-normal* distribution).

6.5 On the Mathematical Structure of Markov Processes

In this chapter and the previous one we have discussed several examples of Markov stochastic processes, which differ in the continuous or discrete nature of the states and time, as summarized in the following table

	States	Times	Equation
Markov chains	Discrete	Discrete	$\mathbf{p}(t+1) = W^T \mathbf{p}(t)$
Master equation	Discrete	Continuous	$\frac{d\mathbf{p}(t)}{dt} = A\mathbf{p}(t)$
Fokker-Planck	Continuous	Continuous	$\frac{\partial p(x,t)}{\partial t} = Lp(x,t)$

where W is the transition probability matrix, A the transition rate matrix, and the operator L is a differential operator in the case of "displacements small for small times" (as explained in Sect. 6.2) where the Fokker-Planck equation holds, or, in general, an integral operator. Note that the linear structure of the equations is a consequence of the Markovian nature of the processes.

6.6 An Application of Stochastic Differential Equations to Climate Studies

We conclude this chapter by discussing an application of the Langevin equations to a (highly simplified) model of long-term climate variations.

6.6.1 *SDEs as Effective Models*

In Chap. 4 we introduced the Langevin equation as a phenomenological description of Brownian motion. In that case, the possibility of decomposing the force acting on the colloidal particle into two parts (the deterministic one given by Stokes' formula and the stochastic one due to molecular collisions) is due to the separation of the characteristic timescales of the two processes.

Let us now consider a deterministic system whose state is a vector $\mathbf{z} = (\mathbf{x}, \mathbf{y})$ consisting of two classes of variables, $\mathbf{x}$ and $\mathbf{y}$, which are slow and fast variables, respectively. We can then write the evolution equations in the form

$$\frac{d\mathbf{x}}{dt} = \mathbf{f}(\mathbf{x}, \mathbf{y}), \quad \frac{d\mathbf{y}}{dt} = \frac{1}{\epsilon}\mathbf{g}(\mathbf{x}, \mathbf{y})$$

where $\epsilon \ll 1$ is the ratio between the characteristic timescale of the fast variables and that of the slow variables.

Suppose we are interested in the behavior of only the slow variables: a natural approach is to seek an "effective" equation for $\mathbf{x}$ alone that somehow takes into account the presence of $\mathbf{y}$. This problem is very common: we can say that almost all interesting problems in science and engineering are characterized by the presence of degrees of freedom with very different timescales. Perhaps the first study of a system with a multiscale time structure is due to Newton, who analyzed the precession of the equinoxes by properly treating the (fast) motion of the Moon.[6] Among the many important problems involving processes with very different timescales, we can mention the problem of protein folding and climate dynamics. While the vibration period of covalent bonds is $O(10^{-15}\,\text{s})$, the folding time of proteins can be on the order of seconds. Similarly, in climate problems, the characteristic timescales of the processes involved range from days (for the atmosphere) to thousands of years (for deep ocean currents and glaciers).

The general solution to this problem is not at all easy; for the slow variables, one can expect (by analogy with the motion of the colloidal particle discussed in Chap. 4) a Langevin equation of the type:

$$\frac{dx_i}{dt} = F_i(\mathbf{x}) + \sum_{j=1}^{N} A_{i,j}\eta_j \tag{6.48}$$

where η is a white noise vector, i.e., each component is Gaussian and we have

$$< \eta_j(t) >= 0, \quad < \eta_j(t)\eta_i(t') >= \delta_{ij}\delta(t-t').$$

Equation (6.48) essentially expresses in mathematical terms what physical intuition suggests: *the fast degrees of freedom renormalize the systematic part and add a white noise.*

If $\epsilon \ll 1$, under very general assumptions, it can be shown that (6.48) is correct; however, the analytical determination of the form of $\mathbf{F}$ and the matrix $\mathbf{A}$, which generally depends on $\mathbf{x}$, is very difficult, and almost always one must resort to ad hoc methods and phenomenological ideas.

As an example of a case in which the construction of an effective Langevin equation is explicitly possible (in a relatively simple way), we can mention the motion of a particle transported by a velocity field:

$$\frac{dx_j}{dt} = u_j(\mathbf{x}, t) + \sqrt{2D}\eta_j$$

where $\mathbf{u}(\mathbf{x}, t)$ is a given field with characteristic length ℓ and typical time τ. The large-scale (L) and long-time (T) behavior (i.e., $L \gg \ell$ and $T \gg \tau$), under very general assumptions, is described by an effective equation of the form:

$$\frac{dx_j}{dt} = \sum_{i=1}^{N} A_{j,i}\eta_i$$

[6] Newton's idea was to replace the periodic motion of the Moon with a ring of equal mass around the Earth.

where the matrix **A** can be explicitly calculated starting from $\mathbf{u}(\mathbf{x}, t)$. In other words, the velocity field over long times has the effect of renormalizing the diffusion coefficient; in the isotropic case, this is particularly evident, and we have

$$\frac{dx_j}{dt} = \sqrt{2D^{eff}}\,\eta_j$$

where D^{eff} depends, often in a non-intuitive way, on the field $\mathbf{u}(\mathbf{x}, t)$.

In Chap. 7 we will explicitly discuss an example of Lagrangian transport in a laminar velocity field.

6.6.2 A Simple Stochastic Model for Climate

After the previous discussion, let us now introduce a very simple model of climate dynamics for very long timescales (on the order of thousands of years). As a slow variable, we take the average temperature $T(t)$ (meaning the average over the entire Earth and over a relatively long time interval, say tens or hundreds of years). From the concentration of the oxygen isotope 18 contained in fossil plankton, it is possible to reconstruct the Earth's temperature over the last million years.[7] An approximately periodic trend is observed with a period of about 10^5 years, with a transition between two values T_1 and T_2 (let's say cold and warm climate, respectively) with $\Delta T = T_2 - T_1 = O(10)$ K (Fig. 6.2).

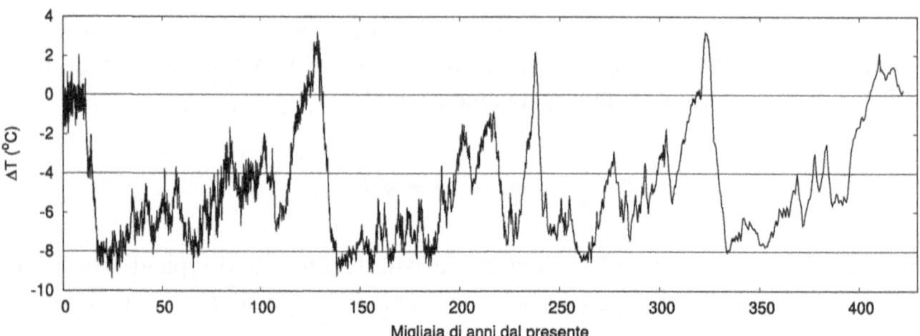

Fig. 6.2 Variation of the Earth's average temperature over the last 450,000 years reconstructed from the Vostok ice core. On the x-axis, thousands of years before present; on the y-axis, the temperature difference with respect to the present. Data from: J. Jouzel, C. Lorius, J.R. Petit, C. Genthon, N.I. Barkov, V.M. Kotlyakov, V.M. Petrov, *Nature* **329**, 403 (1987)

[7] Fossil plankton is deposited at the bottom of the ocean. Since the oxygen isotope 18 evaporates differently from isotope 16 (which is much more common), the ratio between the two concentrations depends on temperature. Given the known relationship between the percentage distribution curve of the isotopes and temperature, by analyzing the different ratio between the oxygen isotopes, it is possible to reconstruct with some precision the temperature of the sea waters.

Obviously, in the evolution of $T(t)$, both slow and fast processes are involved[8] among the fast ones is atmospheric dynamics with its circulation, turbulence, etc. We have an equation of the form:

$$\frac{dT}{dt} = F(T,t) + \sqrt{2D}\eta$$

where the term $F(T,t)$ describes the contribution of the slow dynamics, while the white noise accounts for "fast" physical processes (say, on the order of months or years).

We can write $F(T,t)$ as the sum of two contributions:

$$F(T,t) = In(T,t) - Out(T),$$

$In(T,t)$ is determined by the energy flux coming from the Sun, which depends on time because the Earth's orbit, due to interaction with the Moon and the other planets (mainly the larger ones like Jupiter and Saturn), is not constant; in particular, there are variations in eccentricity and precession of the rotation axis. The temporal evolution of the orbital parameters was accurately determined by Milankovitch with celestial mechanics calculations; the most relevant contribution is given by the eccentricity, which approximately varies periodically with a period of about 10^5 years:

$$In(T,t) = Q\Big(1 + \epsilon cos(\omega t)\Big)$$

where Q is a constant (related to the mean solar constant), $\epsilon = O(10^{-3})$, $\omega = 2\pi/T$ with $T \simeq 10^5$ years.

The term $Out(T)$ consists of two contributions, one rather simple $C\,T^4$ (Stefan-Boltzmann law), and another more delicate one due to the part of the energy coming from the Sun that is reflected: $A(T)In(T,t)$, where $A(T)$ represents the albedo, a numerical coefficient between 0 and 1 that measures the fraction of radiation reflected. The calculation of the albedo is rather complicated since it is determined by many factors, including the fraction of Earth's surface covered by glaciers, deserts, forests, the concentration of particulates in the atmosphere, and so on. For simplicity, we consider the albedo empirically as a decreasing function of Earth's temperature (Fig. 6.3).

We therefore have

$$F(T,t) = Q\Big(1 + \epsilon cos(\omega t)\Big)\Big(1 - A(T)\Big) - C\,T^4. \tag{6.49}$$

[8] Many phenomena are involved with very different time scales, ranging from seconds (microturbulence in the atmosphere and ocean) to months and years (geostrophic vortex structures in the ocean) up to millennia (thermohaline circulation).

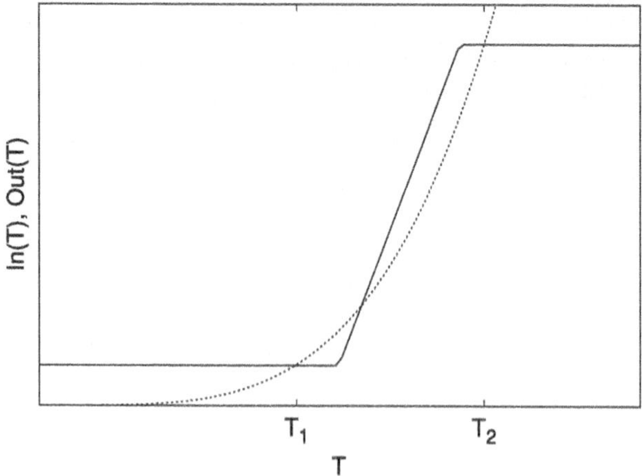

Fig. 6.3 Input term $Q(1 - A(T))$ (solid line) and output term CT^4 (dashed line) in Eq. (6.49) for $\epsilon = 0$. The intersections between the two curves correspond to the fixed points $F = 0$, of which T_1 and T_2 are stable (minima of the potential $V(T)$), while the intermediate point is unstable

Milankovitch, by comparing observations of temperature trends over the last million years and the evolution of the astronomical parameters of Earth's orbit, observed a strong correlation between $T(t)$ and $In(T, t)$. It would therefore seem natural to attribute long-term climate changes to variations in Earth's orbital parameters. The problem is that the variations in the solar constant are actually very small, $\epsilon \simeq O(10^{-3})$, and therefore it is necessary to think of an amplification mechanism.

For $\epsilon = 0$, $F(T)$ has the form

$$F(T) = -\frac{\partial V}{\partial T}$$

where $V(T)$ diverges for $T \to \pm\infty$ and has two minima at T_1 and T_2 with $\Delta T = T_2 - T_1 = O(10)$ Kelvin (see Fig. 6.3). In the absence of eccentricity variations and without white noise, Eq. (6.49) has two stable solutions T_1 and T_2. Introducing the (small) periodic perturbation, since the "potential" $V(T, t)$ at each time always has two minima, $T(t)$ varies only slightly around T_1 or T_2 (depending on the initial condition), but there are no transitions between the two climate states.

Including the stochastic term, it becomes possible to have transitions between the two minima. In the absence of the periodic term, these transitions occur irregularly; in fact, the interval τ between two successive transitions is a random variable, in which the successive jumps are independent events.

The interesting (and somewhat unexpected) result is that there exists a range of values of the noise amplitude D in which transitions between the two states occur in substantial synchrony with the periodic perturbation, that is, the residence time in each of the two minima is about $5\ 10^4$ years. This is the mechanism of *stochastic resonance*, originally introduced precisely for climate models and later studied in very general contexts, from biology to laser physics to engineering.

6.6.3 The Mechanism of Stochastic Resonance

Let us consider a one-dimensional stochastic differential equation of the form

$$dx = -\frac{\partial V_0(x)}{\partial x}dt + \sqrt{2D}dW(t) \tag{6.50}$$

where $V_0(x)$ is a symmetric function with two minima x_-, x_+ and a maximum x_0 (Fig. 6.4), and we denote by ΔV the difference in potential between the maximum and the minimum $\Delta V = V_0(x_0) - V_0(x_-)$. For example,

$$V_0(x) = \frac{1}{4}x^4 - \frac{1}{2}x^2. \tag{6.51}$$

Assume that at the initial instant the system is in the left well; linearizing the problem around x_-, for the variable $z = x - x_-$ we have

$$dz = -\frac{1}{t_C}zdt + \sqrt{2D}dW(t)$$

where $t_C = 1/\sqrt{V_0''(x_-)}$. The previous equation is approximately valid for small values of z, that is, as long as $x(t)$ remains in the left well. After some time, there will be a transition to the right well and then again to the left, and so on. In the limit

$$\frac{\Delta V}{D} \gg 1$$

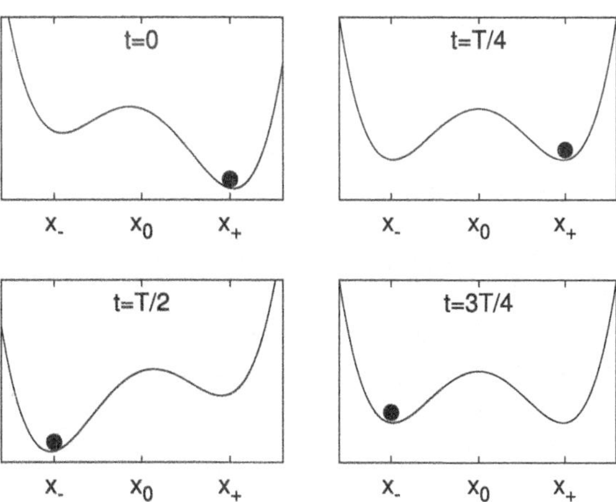

Fig. 6.4 Mechanism of stochastic resonance. The oscillation period of the double well is T. At time $t = 0$ the system has a very low probability of jumping from x_+ to x_-. At time $t = T/2$ the situation is reversed and the probability of not jumping is very low, so the particle will jump into the left well. After a period T, a new cycle begins

the mean jump time is given by Kramers' formula (6.37)

$$\tau \simeq \tau_0 e^{\frac{\Delta V}{D}}$$

with $\tau_0 = \pi/\sqrt{|V_0''(x_-)V_0''(x_0)|}$.

Let us now discuss the case in which the potential has a periodic perturbation (as in the climate dynamics model)

$$V(x,t) = V_0(x) + ax\cos(\omega t) \tag{6.52}$$

with period $T = 2\pi/\omega$ and where a is small enough so that at any time $V(x,t)$ still presents two minima and one maximum (which vary slightly compared to the stationary case).

The potential barrier $\Delta V(t)$ between the minimum and the maximum varies in time between two values $\Delta V_{min} \simeq \Delta V_0 - a$ and $\Delta V_{max} \simeq \Delta V_0 + a$ (to first order in a), and therefore the mean escape time changes over time. If the period T is much greater than the typical time around one of the two minima $t_C = 1/\sqrt{|V_0''(x_\pm)|}$, we can assume we are in "adiabatic" conditions, so that the mean escape time is given by (6.51), that is,

$$\tau(t) \simeq \tau_0(t) e^{\frac{\Delta V(t)}{D}}.$$

The key point is the strong (exponential) dependence of the escape time on ΔV. In an appropriate range of values of D, we will have that $\tau(0) \simeq \tau_0 e^{\Delta V_{max}/D} \gg T/2$ and therefore a transition is unlikely, while $\tau(T/2) \simeq \tau_0 e^{\Delta V_{min}/D} < T/2$ and thus with probability close to one there will be a transition right around the time $t = T/2$. After the transition, the system is in the deep well and, repeating the reasoning, the next jump will occur after another half period.

In Fig. 6.5c the behavior of $x(t)$ is shown under conditions of stochastic resonance: $x(t)$ makes transitions between x_- and x_+ in an approximately periodic way and synchronized with the frequency of the potential perturbation (shown in Fig. 6.5b). As shown in Fig. 6.5a, in the absence of perturbation the jumps instead occur at random intervals. Note that, under conditions of stochastic resonance, if the system misses the opportunity to make the transition at the "right moment" it must wait a period; this implies that the probability density of τ is peaked around $T/2$, $3T/2$, $5T/2, \ldots$.

The qualitative arguments discussed above are confirmed by detailed analytical studies, numerical simulations, and experiments.

The reader might wonder how realistic the mechanism of stochastic resonance is in the real climate. This is not the place for a technical discussion; however, we can mention that the study of realistic models, which include for example the general circulation of the oceans, shows that the probability distribution of the transition time between two climate states (warm/cold) is quantitatively similar to that predicted by the stochastic resonance mechanism.

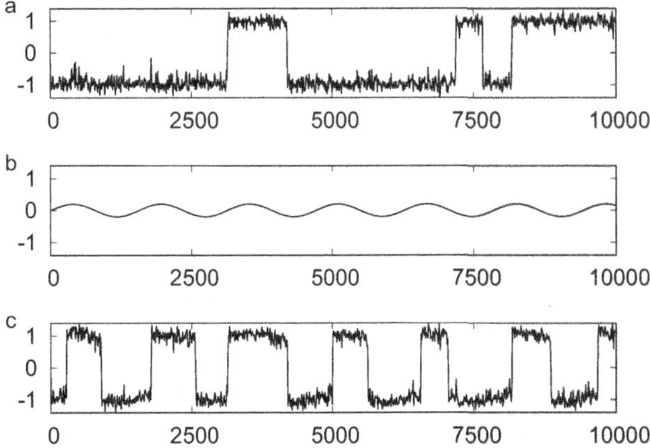

Fig. 6.5 Demonstration of the stochastic resonance mechanism by means of numerical simulations. Graph (**a**) corresponds to the solution of Eq. (6.50) with stationary potential (6.51). In this case, the system jumps from one well to the other at random times determined only by the noise according to (6.51). If we introduce a periodic perturbation in the potential as described in (6.52) and shown in graph (**b**) (with $a = 0.1$), the system jumps from one well to the other synchronized with the frequency of the perturbation, as shown in graph (**c**). Figure taken from R. Benzi, *Nonlin. Processes Geophys.* **17**, 431 (2010)

Exercises

6.1 Given the Langevin equation

$$\frac{dx}{dt} = -\frac{x}{\tau} + c\eta$$

where η is a white noise, i.e., a Gaussian process with

$$< \eta(t) >= 0 \;\; and \;\; < \eta(t)\eta(t') >= \delta(t - t'),$$

show that

$$< x(t)x(0) >=< x^2 > e^{-t/\tau}.$$

6.2 Consider the discrete-time stochastic process

$$x_{t+1} = ax_t + b \sum_{j=0}^{N-1} 2^{-j} z_{t-j}$$

where $|a| < 1$ and the $\{z_j\}$ are i.i.d. Gaussians with zero mean and unit variance; show that in the limit $t \to \infty$ the variable x_t is Gaussian.

6.3 Given the Langevin equation

$$\frac{dx}{dt} = -\frac{dV(x)}{dx} + c\eta$$

where η is a white noise, and

$$V(x) = \frac{x^4}{4} - \frac{x^2}{2}$$

numerically verify the Kramers approximation by calculating the mean value of the jump time between the two minima of $V(x)$ as c varies.

Note There exist numerical methods for the numerical integration of stochastic differential equations that generalize Runge-Kutta type algorithms, see for example R. Mannella and V. Palleschi "Fast and precise algorithm for computer simulation of stochastic differential equations" Phys. Rev. A **40**, 3381 (1989).

Suggested Readings

A classic article on which generations of physicists have learned the fundamental notions about Markov processes and the Fokker-Planck equation:

M.C. Wang, G.E. Uhlenbeck, On the theory of the Brownian motion II. Rev. Mod. Phys. **17**, 323 (1945)

A brief review suitable for physics students:

A.J. McKane, Stochastic processes, in *Encyclopedia of Complexity and System Science*, ed. by R.A. Meyers (Springer, New York, 2009), pp. 8766–8783

Two very comprehensive books:

H. Risken, *The Fokker-Planck Equation. Methods of Solution and Applications* (Springer, Berlin, 1989);

C. Gardiner, *Handbook of Stochastic Methods: for Physics, Chemistry and the Natural Sciences* (Springer, Berlin, 2004)

For a discussion on the use of Langevin equations to describe "complex" systems:

P. Imkeller, J.S. von Storch (eds.), *Stochastic Climate Models* (Birkhäuser, Basel, 2001)

Stochastic resonance was originally introduced to explain long-term climate variations:

R. Benzi, A. Sutera, A. Vulpiani, The mechanism of stochastic resonance. J. Phys. A Math. Gen. **14**, L453 (1981);

R. Benzi, G. Parisi, A. Sutera, A. Vulpiani, Stochastic resonance in climatic change. Tellus **34**, 10 (1982);

For a detailed review of the applications of stochastic resonance in physics, chemistry, biology, and engineering:

L. Gammaitoni, P. Hänggi, P. Jung, F. Marchesoni, Stochastic resonance. Rev. Mod. Phys. **70**, 223 (1998)

Chapter 7
Probability and Deterministic Chaotic Systems

As we have seen in previous chapters, the use of the theory of probability in physical problems originates with problems in statistical mechanics, that is, with the reduction to a few macroscopic variables of a system composed of a large number of microscopic components. The example of Brownian motion is, in this regard, illuminating: the treatment by Einstein and Langevin reduces the dynamics of the system composed of the pollen grain and an enormous number (on the order of 10^{23}) of molecules to the macroscopic motion of the grain in the presence of noise. The use of probabilistic methods is in these cases justified by the fact that the superposition of many causes (the collisions of the molecules) leads to an effect that can be treated in statistical terms.

For deterministic chaotic systems, the use of probabilistic description is not motivated by the large number of degrees of freedom (in fact, there are chaotic systems even with a single degree of freedom), but by the erratic, pseudo-random nature typical of chaotic trajectories.

In this chapter, we will show how the theory of probability can be very useful for the study of chaotic dynamical systems by identifying the analogies and differences between deterministic chaotic systems and Markov processes.

7.1 The Discovery of Deterministic Chaos

The study of chaotic dynamics in deterministic systems arises from the observation that many physical systems, while governed by deterministic laws, display irregular and unpredictable behavior.

Without invoking complex dynamics such as the motion of the Earth's atmosphere, it is sufficient to consider (and, if desired, realize in the laboratory) very simple chaotic systems. Even a pendulum, the prototype of regular motion in physics, if forced periodically (for example, by oscillating the point of suspension

G. Boffetta, A. Vulpiani, *Probability in Physics*, UNITEXT for Physics,
https://doi.org/10.1007/978-3-032-10407-6_7

vertically) can display a very complicated chaotic motion. This motion is *in fact* unpredictable, in the sense that even a very small perturbation in the initial condition can generate a completely different trajectory. This *sensitive dependence on initial conditions* is poetically summarized by the so-called *butterfly effect* introduced by the title of a lecture given by the meteorologist Lorenz: "*Can the flap of a butterfly's wings in Brazil set off a tornado in Texas?*"

The modern study of chaotic systems is traced back to an important paper published by Lorenz in 1963 in the *Journal of Atmospheric Sciences*.[1] In this work, starting from the equations governing the motion of a fluid heated from below (in his case, the atmosphere), Lorenz derives, through major simplifications, a model composed of 3 differential equations that today bears his name

$$\begin{cases} \dot{X} = \sigma(Y - X) \\ \dot{Y} = X(r - Z) - Y \\ \dot{Z} = XY - bZ \end{cases} \tag{7.1}$$

In this model, the variables (X, Y, Z) represent, respectively, the intensity of the convective motion, the temperature difference between ascending and descending currents, and the distortion of the vertical temperature profile from linearity. The parameter σ is the ratio between viscosity and diffusivity of the fluid (Prandtl number), b is related to the geometric aspect of the convective cell, while r represents the relative Rayleigh number, proportional to the temperature difference that drives the system.

A trivial solution of system (7.1) is obviously the origin $X = Y = Z = 0$, which physically corresponds to the conductive regime in which the fluid is at rest. However, this solution becomes unstable as the parameter r increases, and the study of the dynamics must be addressed through numerical simulations. The numerical simulation of (7.1) with parameter values $\sigma = 10$, $b = 8/3$, and $r = 28$ allowed Lorenz to discover a new regime characterized by deterministic chaos, defined by the properties:

1. *asymptotic aperiodic behavior*, in the sense that the trajectory does not converge to a fixed point or a periodic orbit;
2. *sensitive dependence on initial conditions*, that is, two trajectories starting from infinitesimally close initial conditions separate exponentially in time.[2]

To avoid misunderstandings, let us emphasize that the chaotic systems described by these properties are completely deterministic, as in example (7.1). There is nothing intrinsically stochastic, and therefore two realizations that start *exactly* from the same initial condition will retrace the same trajectory. In any physical problem,

---

[1] Although in reality the first example of chaotic motion was discovered by Poincaré in the study of the three-body problem.

[2] This property seems to have been discovered partly by chance by Lorenz as a consequence of the limited precision of the initial condition entered into the computer.

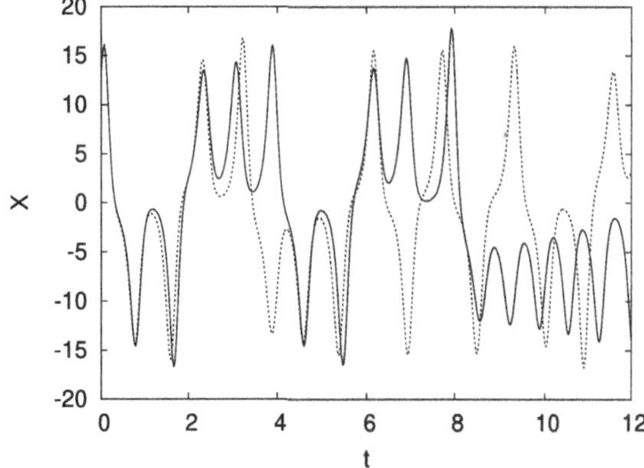

Fig. 7.1 Time evolution of the variable X of the Lorenz system for two initial conditions that initially differ by 10^{-4}

however, we always deal with finite resolution (of the measurement, or of the representation of real numbers on a computer), and therefore an uncertainty in the initial condition is always present. In a chaotic system, this initial uncertainty rapidly amplifies (exponentially) in time and quickly leads to the impossibility of making predictions about the future state of the system. In this sense, chaotic systems behave, over long times, essentially like stochastic systems, and it is therefore useful to study them using probabilistic techniques.

As an example, let us consider two trajectories of the system (7.1) that start from very close initial conditions. Figure 7.1, which shows the X component of the two trajectories, clearly displays both properties of chaotic systems discussed earlier. Each of the trajectories jumps from positive to negative values in an irregular way, without apparent periodicity, preventing the possibility of predicting the future state of the system based on its past behavior. The comparison between the two trajectories, on the other hand, shows the second property of chaotic systems: although the two realizations are initially indistinguishable (the initial difference is in fact 10^{-4}), they quickly separate over time, generating two essentially independent trajectories.

To understand the origin of the chaotic behavior of the system (7.1), let us consider, following Lorenz, a discrete-time representation of the trajectory. Starting from the trajectory of the variable Z, we define m_n as the value taken by the n-th local maximum. By plotting on a graph (called a return map) the value of the maximum m_{n+1} as a function of the previous maximum m_n, we obtain Fig. 7.2. We observe that the points with coordinates (m_n, m_{n+1}) are arranged along a regular curve (unlike what would be obtained with a stochastic system), which can therefore be described by a functional relationship

$$m_{n+1} = f(m_n). \tag{7.2}$$

The advantage of studying the return map (7.2) is that, although it is simpler than the original system (7.1), it essentially contains the information of the continuous

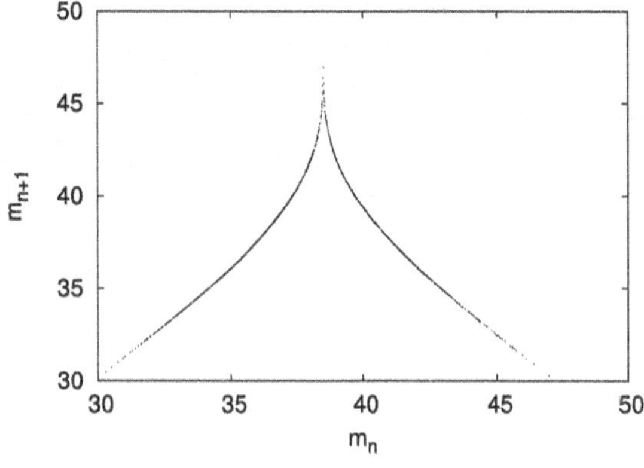

Fig. 7.2 Return map for the variable Z of the Lorenz system obtained by plotting the value taken by a maximum of Z as a function of the value of the previous maximum

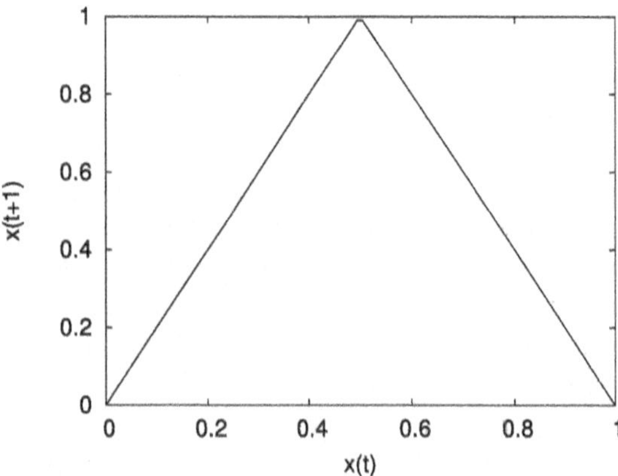

Fig. 7.3 Graph of the tent map

trajectory. For example, in the case of a periodic system, we would have $m_{n+1} = m_n$ and therefore a return map consisting of a single point. The analysis is further simplified if we consider, following Lorenz, an idealization of the map (7.2) given by the tent map

$$x(t+1) = f_T(x(t)) = \begin{cases} 2x(t) & \text{if} \quad x(t) < 1/2 \\ 2 - 2x(t) & \text{if} \quad x(t) \geq 1/2 \end{cases} \tag{7.3}$$

with integer t and represented in Fig. 7.3 (in the transition from the map in Fig.7.2 to the tent map, the variable has been rescaled to take values between 0 and 1). The advantage of using the tent map is that, thanks to its piecewise linearity, it is possible to analytically study some properties and thus obtain insights into the behavior of chaotic systems like the original Lorenz system.

Recall that a number $x \in [0, 1)$ can be represented by the binary sequence $\{\alpha_j\}_{j \in \mathbb{N}}$ with $\alpha_j = \{0, 1\}$ as

$$x = \sum_{j \in \mathbb{N}} \alpha_j 2^{-j}$$

which we denote with the compact notation $x = [\alpha_1, \alpha_2, \ldots]$. Now consider an initial condition for the tent map

$$x(0) = [\alpha_1, \alpha_2, \alpha_3, \ldots]$$

and its iterate $x(1) = f_T(x(0))$. If $\alpha_1 = 0$, then $x(0) < 1/2$ and we have

$$x(1) = [\alpha_2, \alpha_3, \alpha_4, \ldots] \tag{7.4}$$

whereas if $\alpha_1 = 1$, $x(0) \geq 1/2$ and, introducing the "negation" operator N (such that $N0 = 1$ and $N1 = 0$), we have

$$x(1) = [N\alpha_2, N\alpha_3, N\alpha_4, \ldots] \tag{7.5}$$

For example, taking $x(0) = [1, 0, 1, 0, 1, 0 \ldots] = 2/3$ we obtain $x(1) = [N0, N1, N0, N1 \ldots] = [1, 0, 1, 0, 1, 0 \ldots] = 2/3$ which is a fixed point of the map.

Defining $N^0 = I$ (identity operator), the two relations (7.4) and (7.5) can be combined as

$$x(1) = [N^{\alpha_1}\alpha_2, N^{\alpha_1}\alpha_3, N^{\alpha_1}\alpha_4, \ldots].$$

By iterating the procedure, we obtain the explicit solution after n steps

$$x(n) = [N^{\alpha_1+\alpha_2+\ldots+\alpha_n}\alpha_{n+1}, N^{\alpha_1+\alpha_2+\ldots+\alpha_n}\alpha_{n+2}, \ldots]. \tag{7.6}$$

From (7.6) it is immediately clear that "almost all" trajectories of the tent map are non-periodic. In fact, by definition, only those trajectories that start from an initial condition given by a rational number, i.e., representable as

$$x(0) = [\alpha_1, \ldots, \alpha_p, \beta_1, \ldots, \beta_q, \beta_1, \ldots, \beta_q, \ldots]$$

(where the sequence $(\beta_1, \ldots, \beta_q)$ repeats indefinitely) with $p, q \in \mathbb{N}$, will be periodic, and the resulting trajectory, after a transient of p steps, will be confined to a cycle of period q. On the other hand, all trajectories that start from an irrational initial condition, which are infinitely more numerous (since the rational numbers form a countable set of measure zero within the real numbers), will be non-periodic.

The second property of chaotic systems (sensitive dependence on initial conditions) can also be easily verified for the tent map using the solution (7.6). Let us consider two trajectories x and y that are initially close, in the sense that their initial conditions differ only starting from the m-th digit of the binary representation. As an example, let us take $x(0) = [0, 1, 1, 0, 1, 0, 1, 1, 0, 0, 1, 0, 1 \ldots]$ and $y(0) = [0, 1, 1, 0, 1, 0, 1, 1, 0, 0, 1, 0, 0 \ldots]$ ($m = 13$) and thus $\delta(0) \equiv |x(0) - y(0)| = 2^{-13} \simeq 0.0001$. After one step we have $x(1) = [1, 1, 0, 1, 0, 1, 1, 0, 0, 1, 0, 1 \ldots]$ and $y(1) = [1, 1, 0, 1, 0, 1, 1, 0, 0, 1, 0, 0 \ldots]$ and therefore the difference will be

$\delta(1) \simeq 0.0002$. The distance between the trajectories thus grows exponentially in time as $\delta(n) = \delta(0)2^n$, and after $n = m$ steps we have $\delta(n) = O(1)$.

This elementary example explicitly shows that despite the deterministic (and simple!) nature of the tent map, predicting the behavior of the trajectory at long times is impossible because an arbitrarily small error in the initial condition quickly leads to a macroscopic uncertainty. We also observe that the time for this amplification of uncertainty (called the *predictability time*) is an intrinsic characteristic of the system that depends little on the initial uncertainty. In fact, if we denote by δ_{max} the maximum acceptable uncertainty in the knowledge of the trajectory, which in the case of the tent map will be $\delta_{max} = O(1)$, we have that the predictability time (measured in number of iterations) is given by the relation $T_{max} = \log_2(\delta_{max}/\delta(0))$ and therefore increases very slowly, logarithmically, as the precision in the initial condition increases.

7.2 The Probabilistic Approach to Chaotic Dynamical Systems

As we saw in the previous section, deterministic chaos is characterized by the practical impossibility of predicting the long-term behavior of a single trajectory due to the exponential separation of nearby trajectories. A natural question is to ask about the evolution of a set of trajectories starting from different initial conditions. Let us therefore consider a set of trajectories distributed in phase space Ω ($\Omega = [0, 1]$ in the case of the tent map) according to an initial probability density $\rho_0(x)$ defined by the fact that $\rho_0(x)dx$ represents the probability of having an initial condition (i.e., the fraction of initial conditions) in the interval $[x, x + dx]$.

The time evolution of the trajectories under a generic map $f : \Omega \to \Omega$

$$x(t+1) = f(x(t)) \tag{7.7}$$

induces the evolution of the density over time in such a way that $\rho_t(x)dx$ represents the probability of finding the trajectory in $[x, x + dx]$ at time t. An analogous and more general concept is that of a *measure* defined from $\rho_t(x)$ as the probability $\mu_t(A)$ of finding the trajectory in a given set $A \subset \Omega$

$$\mu_t(A) = \int_A \rho_t(x)dx.$$

Being a probability density, $\rho_t(x)$ must satisfy (at every time) the normalization condition

$$\int_\Omega \rho_t(x)dx = 1 \tag{7.8}$$

and therefore $\mu_t(\Omega) = 1$.

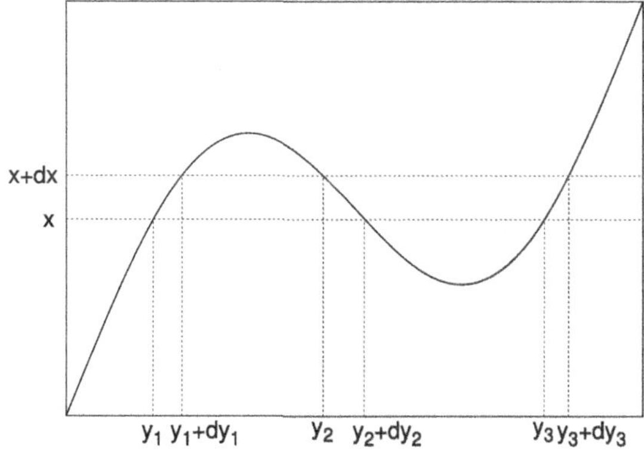

Fig. 7.4 Graphical representation of the evolution of the probability density according to the Perron-Frobenius equation

Since the evolution (7.7) is deterministic, the probability density of finding the trajectory at a point x at time $t+1$ is determined by the probability densities of having the trajectory at the points y_k at time t such that $f(y_k) = x$. The equation for the evolution of the probability density, called the **Perron-Frobenius** (PF) equation, is written as

$$\rho_{t+1}(x) = \int_{\Omega} \rho_t(y)\delta[x - f(y)]dy = \sum_k \frac{\rho_t(y_k)}{|f'(y_k)|}. \tag{7.9}$$

The meaning of this equation is quite clear by noting that

$$P(x(t+1) \in [x, x+dx]) = \rho_{t+1}(x)dx =$$
$$\sum_k P(x(t) \in [y_k, y_k + dy_k]) = \sum_k \rho_t(y_k)dy_k$$

and since $dy_k = dx/|f'(y_k)|$ it follows (7.9), see Fig. 7.4.

Note that (7.9) preserves the normalization of the density (7.8) and therefore represents the conservation of probability under the action of the map (7.7).

It can be shown, under general conditions, that in the case of chaotic systems, the evolution according to the PF equation quickly (exponentially in time) brings any initial probability density $\rho_0(x)$ to the **invariant density** $\rho(x) \equiv \lim_{t\to\infty} \rho_t(x)$, which is invariant under evolution according to (7.9)

$$\rho(x) = \int_{\Omega} \rho(y)\delta[x - f(y)]dy \tag{7.10}$$

and is therefore a fixed point of the PF dynamics. Thus, in chaotic systems, for which the dynamics of a single trajectory is non-periodic and unpredictable, the evolution of the density of trajectories is instead regular and converges to an invariant function, independently of $\rho_0(x)$. It is interesting to note that in the case of non-chaotic systems (for example, with periodic trajectories), the evolution of $\rho_t(x)$ depends

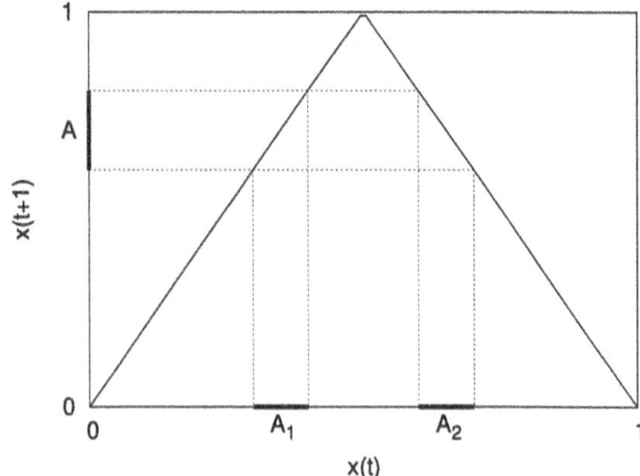

Fig. 7.5 Graphical method for determining the invariant measure of the tent map. In this case, the length of the set A is equal to the sum of the lengths of the preimages A_1 and A_2, and therefore the invariant measure of a set is given by its length

on $\rho_0(x)$ for arbitrarily long times, that is, in this case the PF equation does not forget the initial condition.

As an example, we show that the invariant density for the tent map (7.3) is the constant function $\rho(x) = 1$. Indeed, we have

$$\rho(x) = \int_0^1 dy\delta[x - f(y)] = \int_0^{1/2} dy\delta[x - 2y] + \int_{1/2}^1 dy\delta[x - 2 + 2y]. \quad (7.11)$$

Now, using the well-known property $\delta(\lambda x) = \delta(x)/|\lambda|$ and noting that both integrals in (7.11) are equal to 1/2, we obtain $\rho(x) = 1$, which is therefore a solution of (7.10). In this case, the invariant measure is nothing but the length of the interval (see Fig. 7.5).

A less trivial example of an invariant density is found by considering the so-called **logistic map** defined by

$$x(t + 1) = rx(t)(1 - x(t)) \quad (7.12)$$

where $0 \leq r \leq 4$ so that the variable x remains in the phase space $\Omega = [0, 1)$.

The map (7.12) finds applications in many fields, for example in the study of population dynamics where $x(t)$ represents the number of individuals (appropriately normalized) in a population. For small values of $x(t)$, (7.12) is approximated by the linear equation $x(t + 1) = rx(t)$ (in which r represents the reproductive power of the population) which has the solution $x(t) = r^t x(0)$. If $r < 1$, under the evolution of (7.12) the population will go extinct, that is, $\lim_{t\to\infty} x(t) = 0$, regardless of the initial condition (since each individual is "replaced" by a fraction $r < 1$ of offspring and therefore the population dies out exponentially fast).

When $r > 1$, the linear approximation instead predicts a *population explosion*, i.e., exponential growth of the population. In reality, as x increases, the effective reproductive power is reduced to $r(1 - x)$ and therefore the growth slows down. What makes the logistic map (7.12) interesting is that it exhibits an unexpectedly rich dynamics as the parameter r varies. For $1 < r < 3$ the population reaches an equilibrium state $x = \frac{r-1}{r} > 0$ that does not change over time (and whose value depends on r). For $r > 3$, however, a stationary state is not reached, but the evolution of x can lead to a periodic cycle (in which the population varies cyclically) or even to chaotic dynamics.

The existence of chaotic trajectories for (7.12) can be easily demonstrated in the limiting case $r = 4$ by exploiting the equivalence of the logistic map with the tent map (7.3).

Postponing the details to specialized texts, let us recall that two one-dimensional maps of the type $x(t+1) = f(x(t))$ and $y(t+1) = g(y(t))$ are *topologically conjugate* if there exists an invertible change of variables $y = h(x)$ such that trajectories $x(t)$ are mapped into trajectories $y(t)$ (and vice versa by the inverse transformation h^{-1}). Topological conjugacy between two maps means that they are nothing but two different representations (obtained by a change of variables) of the same system.

Two maps topologically conjugate by the transformation $y = h(x)$ have their respective invariant densities $\rho_x(x)$ and $\rho_y(y)$ related by the formula obtained in Chap. 2 for (monotonic) variable transformations, namely

$$\rho_y(y) = \rho_x(x)/|h'(x)|.$$

It can be directly verified that the tent map in the variable y and the logistic map (7.12) at $r = 4$ are conjugate by the transformation $x = h^{-1}(y) = (1 - \cos(\pi y))/2$. Since

$$\frac{dh^{-1}(y)}{dy} = \frac{\pi}{2}\sin(\pi y) = \pi\sqrt{x(1-x)}$$

and $\rho_y(y) = 1$, we obtain the invariant density for the logistic map at $r = 4$

$$\rho_x(x) = \frac{1}{\pi\sqrt{x(1-x)}}. \tag{7.13}$$

Figure 7.6 shows the evolution of the distribution of 10^5 trajectories starting in a small neighborhood of $x_0 = 0.4$. The distribution, initially peaked around the value x_0, after a few steps (at $t = 8$ in the figure) spreads out to cover a wider interval. At very long times, the distribution of the trajectories converges to the invariant density (7.13) independently of the initial distribution.

7.2.1 Ergodicity

An important property of dynamical systems associated with the invariant density is that of **ergodicity**. The study of ergodic properties originated with Boltzmann

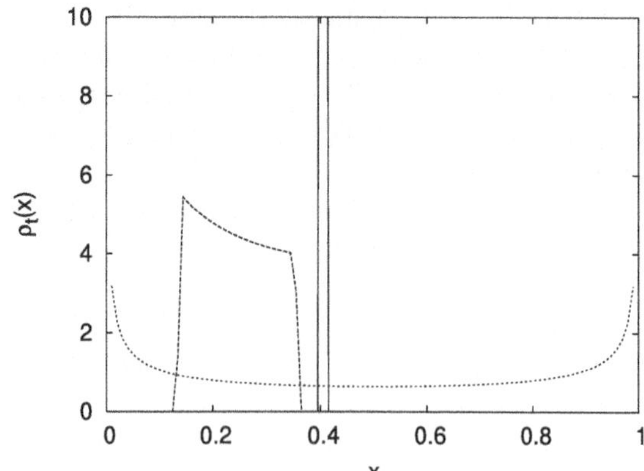

Fig. 7.6 Histogram of 10^5 trajectories of the logistic map $x(t+1) = 4x(t)(1 - x(t))$ with initial conditions in a small neighborhood of $x_0 = 0.4$ calculated at $t = 0$ (solid line), $t = 8$ (dashed line), and at long times (dotted line)

and with the need to identify theoretical expectation values (obtained from an ensemble average) with experimental measurements (which are time averages over fast microscopic variables). Since then, ergodic theory has evolved into an autonomous line of research that can be seen as a branch of measure theory. Referring the interested reader to specialized texts for further details, we recall here that a dynamical system (in our case a one-dimensional map) is ergodic if, for every integrable test function $\phi(x)$, the average calculated with the invariant density is equivalent to the time average along a trajectory, that is,

$$\bar{\phi} \equiv \int \phi(x)\rho(x)dx = \lim_{T\to\infty} \frac{1}{T} \sum_{k=0}^{T-1} \phi(x(k)) \equiv \langle \phi \rangle$$

which therefore does not depend on the initial condition $x(0)$.

The proof of the ergodicity, or lack thereof, of a dynamical system is not generally simple. An important result in this sense, due to Birkhoff and von Neumann, is that a dynamical system is ergodic if and only if for every invariant set A (for which, that is, if $x \in A$ then $f(x) \in A$ as well) we have $\mu(A) = 1$ or $\mu(A) = 0$. In other words, and in agreement with intuition, for an ergodic system there do not exist invariant subsets $A \subset \Omega$ of nonzero measure and different from 1 (otherwise, the trajectories would be confined to them, thus violating ergodicity).

It is important to note that the condition of ergodicity is not sufficient to guarantee the chaoticity of the system. In fact, consider the example given by the translation map

$$x(n+1) = (x(n) + a) \bmod 1$$

defined again on the unit interval $\Omega = [0, 1)$. If a is irrational, any trajectory will uniformly cover Ω, thus reproducing the invariant density (which is obviously $\rho(x) = 1$) (see Exercises). On the other hand, the system is clearly not chaotic (but quasi-periodic), and in fact the distance between two trajectories remains constant over time. Note also that, precisely because of the non-chaotic nature of the system, the initial distribution of trajectories $\rho_0(x)$ will evolve over time simply by translating by a at each iteration and therefore will not converge to the invariant distribution, unlike the example of the tent map.

Therefore, the ergodicity of a system does not guarantee convergence to the invariant density starting from a generic $\rho_0(x)$. This convergence is obtained under the stronger condition that the system is **mixing**. A dynamical system (in our case a one-dimensional map) is said to be mixing if, given any pair of subsets $A, B \subset \Omega$, we have

$$\lim_{t\to\infty} \mu(A \cap f^t(B)) = \mu(A)\mu(B) \tag{7.14}$$

where $f^t(B)$ denotes the set obtained by evolving the points of set B for a time t. The definition (7.14) is in agreement with the intuitive idea of a system that "mixes" the trajectories: the probability that a trajectory starts from a point in B and is found at long times in A is simply given by the product of the probabilities of being in A and in B. The measure (and the corresponding density) that satisfies condition (7.14) is the invariant measure of the system.

The mixing condition is, as we have said, stronger than the condition of ergodicity, in the sense that the latter is implied by the former, as can be easily seen. In fact, consider a mixing system and let $A \subset \Omega$ be an invariant set. We will then have $f^t(A) = A$, which, substituted into (7.14), leads to the condition

$$\mu(A) = \mu^2(A) \tag{7.15}$$

The solutions of (7.15) are $\mu(A) = 1$ and $\mu(A) = 0$, and therefore ergodicity is demonstrated.

The property of mixing is very strong in that it guarantees convergence to an invariant measure (which is ergodic, as discussed above), in the sense that the evolution under (7.9) leads to the density $\rho(x)$ independently of the initial density $\rho_0(x)$, and this convergence occurs exponentially fast in time.

It may be useful, as an example, to see how an initial density converges to the invariant density in the case of the tent map (7.3). The evolution under the Perron-Frobenius equation in this case is given, according to (7.9), by

$$\rho_{t+1}(x) = \frac{1}{2}\rho_t\left(\frac{x}{2}\right) + \frac{1}{2}\rho_t\left(1 - \frac{x}{2}\right). \tag{7.16}$$

The explicit evolution can be easily followed if we start from a polynomial initial density. For example, in the case of a parabola

$$\rho_0(x) = a_0x^2 - 2a_0x + c_0$$

where the choice of the coefficient of the linear term is for computational convenience and, for normalization, must be $-2a_0/3 + c_0 = 1$. Under the action of

(7.16), the density after one step of the map becomes $\rho_1(x) = a_1 x^2 - 2a_1 x + c_1$ with

$$a_1 = \frac{1}{4} a_0 \tag{7.17}$$
$$c_1 = -\frac{1}{2} a_0 + c_0.$$

It is therefore an elementary exercise to calculate that the density, after n steps, becomes

$$\rho_n(x) = \left(\frac{1}{4}\right)^n a_0 x^2 - 2\left(\frac{1}{4}\right)^n a_0 x - \frac{2a_0}{3}\left[1 - \left(\frac{1}{4}\right)^n\right] + c_0$$

and thus, asymptotically,

$$\rho(x) = \lim_{n\to\infty} \rho_n(x) = -\frac{2a_0}{3} + c_0 = 1$$

which is the invariant density of the tent map.

The usefulness of this example, which can be extended to the case of initial densities of more complex form, is that it shows that the convergence to the invariant density is exponentially fast in time. This property is common to many mixing systems (of which the tent map is a simple example), and therefore we can think of mixing systems as systems that "quickly forget" the initial condition, in the sense that, whatever the initial condition of the trajectory, the probability of finding it after a certain time t in a set $A \in \Omega$ converges to $\mu(A)$ exponentially in t.

7.3 Chaotic Systems and Markov Chains

The exponentially fast convergence to the invariant density in mixing systems suggests an analogy between chaotic mixing systems and Markov processes. In this section, we will show how this analogy is actually deep and can be formalized for a special class of chaotic systems.

Let us again consider the case of a one-dimensional map on the unit interval $x(t+1) = f(x(t))$ with $x \in \Omega = [0, 1)$ and take a partition of Ω into N distinct intervals $B_j = [b_{j-1}, b_j]$ with $b_j > b_{j-1}$, $b_0 = 0$ and $b_N = 1$. This partition defines a "coarse-grained" symbolic description of the trajectory, in the sense that the sequence $x(0), x(1), x(2), \ldots$ is mapped onto the sequence of integers $i(0), i(1), i(2), \ldots$ where $i(t) = k$ if $x(t) \in B_k$.

We then introduce the $N \times N$ matrix

$$W_{ij} = \frac{\mu_L(f^{-1}(B_i) \cap B_j)}{\mu_L(B_j)} \tag{7.18}$$

where μ_L denotes the Lebesgue measure. We can identify $p_j = \mu_L(B_j)$ with the probability of finding the trajectory $x(t) \in B_j$ and $p(i, j) = \mu_L(f^{-1}(B_i) \cap B_j)$ with the probability that $x(t-1) \in B_j$ and $x(t) \in B_i$. Therefore, we have that $W_{i,j} = p(i, j)/p(j) = p(i|j)$ is the probability of having $x(t) \in B_i$ knowing that $x(t-1) \in B_j$. Note that the normalization condition $\sum_{i=0}^{N} W_{i,j} = 1$ holds, and therefore the matrix $\mathbb{W}$ can be thought of as a transition matrix for a Markov process. Obviously, for this identification to be meaningful, it is necessary that the evolution according to the Markov chain defined by the partition B_i reproduces the dynamics of the original system. More precisely, we can ask whether the stationary distribution π_i solution of $\pi = \pi \cdot \mathbb{W}$ has any relation with the invariant density $\rho(x)$ solution of (7.10).

A rigorous answer to this question is possible in some cases. In particular, if the map is *expanding* ($|df(x)/dx| > 1$ for every $x \in \Omega$), then π_i converges to the invariant density $\rho(x)$ in the sense that for $N \to \infty$

$$\pi_i \to \int_{B_i} \rho(x)dx.$$

There is an entire class of maps, characterized by being expanding, piecewise linear, and with partitions called Markov partitions[3] for which the Markov chain defined by (7.18) generates exactly the invariant density even for finite N. In this case, the symbolic representation of the dynamical system is indeed coarse-grained.

An example of this class is again provided by the tent map (7.3). Let us consider, for example, the partition with $N = 2$, $B_1 = [0, 1/2)$ and $B_2 = [1/2, 1)$ (Fig. 7.6). From (7.18) we obviously have

$$\mathbb{W} = \begin{pmatrix} 1/2 & 1/2 \\ 1/2 & 1/2 \end{pmatrix}$$

which leads to the stationary distribution $\pi_1 = \pi_2 = 1/2$ and therefore $\pi_i = \mu(B_i)$, that is, $\rho(x) = 1$ (see also Exercise 7.4).

7.4 Transport and Diffusion in Fluids

An important example at the intersection between dynamical systems and stochastic processes is given by transport in fluids, a problem that finds applications in various fields, from the study of dispersion in the atmosphere and ocean to effective mixing in solutions.

One possible approach to the problem is the *Lagrangian* one, in which the trajectories of tracer particles are followed (for example, aerosol particles in the

---

[3] A partition B_i is a Markov partition when $f(B_i) \cap B_j \neq 0$ if and only if $B_j \subset f(B_i)$.

atmosphere or microorganisms in the sea) transported by the velocity field $\mathbf{u}(\mathbf{x}, t)$. The fundamental assumption is that these particles are *passive*, meaning that their motion does not modify the underlying flow. A second assumption we will make, which holds in the limit of very small particles, is that the trajectory follows the fluid elements, i.e., the particles move with the local velocity of the fluid (which is not true for particles of finite size, for which inertial effects must be taken into account). We will also consider the case, relevant in most applications, of incompressible flow for which $\nabla \cdot \mathbf{u} = 0$. Taking into account the thermal noise present in the fluid, the trajectory $\mathbf{X}(t)$ of a particle transported by the velocity field is described by a stochastic differential equation (6.40) of the form

$$d\mathbf{X}(t) = \mathbf{u}(\mathbf{X}(t), t)dt + \sqrt{2D}\, d\mathbf{W}(t) \tag{7.19}$$

where D represents the molecular diffusion coefficient.

An alternative approach to the problem is the so-called *Eulerian* one, in which the evolution of the tracer density $\theta(\mathbf{x}, t)$ is followed at every point in space. Apart from a normalization factor, $\theta(\mathbf{x}, t)$ represents the probability of finding a particle at point $\mathbf{x}$ at time t, and therefore the equation for the evolution of the density is given by the Fokker-Planck equation associated with (7.19)

$$\partial_t \theta + \mathbf{u} \cdot \nabla \theta = D \nabla^2 \theta. \tag{7.20}$$

In many applications, the diffusion coefficient D is extremely small and therefore we can neglect the last term in (7.19), which thus becomes an ordinary differential equation

$$\frac{d\mathbf{X}}{dt} = \mathbf{u}(\mathbf{X}(t), t)$$

which formally defines a deterministic system that can generate chaotic trajectories even in relatively simple velocity fields. A very famous example is the stationary flow known as ABC (from the initials of Arnold, Beltrami, and Childress) given by

$$\mathbf{u}(\mathbf{x}) = (A \sin z + C \cos y,\ B \sin x + A \cos z,\ C \sin y + B \cos x)$$

proposed by Arnold in 1965 as a flow that generates chaotic trajectories, studied numerically by Henon in 1966.

In the presence of chaotic trajectories, it becomes natural, following the reasoning described in this chapter, to treat the problem of diffusion in fluids in probabilistic terms. Stochasticity, removed at small scales by assuming $D = 0$, re-enters, so to speak, through the door with the probabilistic description of large-scale diffusion. Let us consider, in fact, the mean square displacement of a set of particles starting from $\mathbf{X}(0) = 0$. We have

$$\frac{d}{dt}\langle X^2(t)\rangle = 2\langle \mathbf{X}(t) \cdot \mathbf{V}(t)\rangle = 2\int_0^t \langle \mathbf{V}(s) \cdot \mathbf{V}(t)\rangle ds = 2\int_0^t C(s)ds \tag{7.21}$$

where, for simplicity of notation, we have introduced the Lagrangian velocity $\mathbf{V}(t) = \mathbf{u}(\mathbf{X}(t), t)$ and the velocity autocorrelation function $C(t) = \langle \mathbf{V}(0) \cdot \mathbf{V}(t) \rangle$. This defines the Lagrangian correlation time

$$\tau_L = \frac{1}{\langle V^2 \rangle} \int_0^\infty C(t) dt$$

which gives a measure of the memory of the Lagrangian dynamics. For $t \ll \tau_L$, the correlation function in (7.21) is approximately $\langle V^2 \rangle$ and the particle moves ballistically with $\langle X^2(t) \rangle = \langle V^2 \rangle t^2$.

If the autocorrelation decays sufficiently rapidly to make the Lagrangian time τ_L finite, for $t \gg \tau_L$ the integral in (7.21) converges and therefore for long times we have

$$\langle X^2(t) \rangle = 2 \langle V^2 \rangle \tau_L t \tag{7.22}$$

that is, a diffusive motion with diffusion coefficient $D = \langle V^2 \rangle \tau_L$ (a result due to Taylor in 1921).

We note the analogy between the result (7.22) and the Langevin derivation in Sect. 4.3. In fact, the fundamental ingredient for having an asymptotic diffusive motion is the finiteness of the temporal correlations of the Lagrangian velocity, regardless of whether the decorrelation is due to a stochastic process or to a deterministic chaotic process. In fact, we will see in Chap. 8 that if the Lagrangian correlation time τ_L diverges, we have an anomalous super-diffusive behavior.

In Eulerian terms, the argument described above is reflected in the fact that for sufficiently long times, (7.20) can be rewritten as a diffusion equation

$$\partial_t \theta = D^E_{ij} \partial_{x_i} \partial_{x_j} \theta$$

where the effective diffusion tensor $D^E_{ij} = \lim_{t \to \infty} \frac{1}{2t} \langle X_i(t) X_j(t) \rangle$ parametrizes the effects of transport and is typically much larger than the molecular coefficient. The problem thus becomes that of calculating D^E_{ij} starting from the velocity field $\mathbf{u}(\mathbf{x}, t)$, a task that is generally very difficult but can be carried out explicitly in the case of simple flows.

As an example, let us consider the problem of transport in an array of alternating vortices of amplitude L, a so-called *cellular* flow, schematically represented in Fig. 7.7 and described by the stream function

$$\psi(x, y) = \psi_0 \sin(kx) \sin(ky)$$

with $k = 2\pi/L$, from which the incompressible velocity field is given by $\mathbf{u}(x, y) = (\partial_y \psi, -\partial_x \psi) = (k\psi_0 \sin(kx) \cos(ky), -k\psi_0 \cos(kx) \sin(ky))$.

This particular flow is suggested by the study of convection problems, in a particular configuration (called *Rayleigh-Bénard*) in which a thin layer of fluid is heated from below between two planes. For sufficiently small temperature

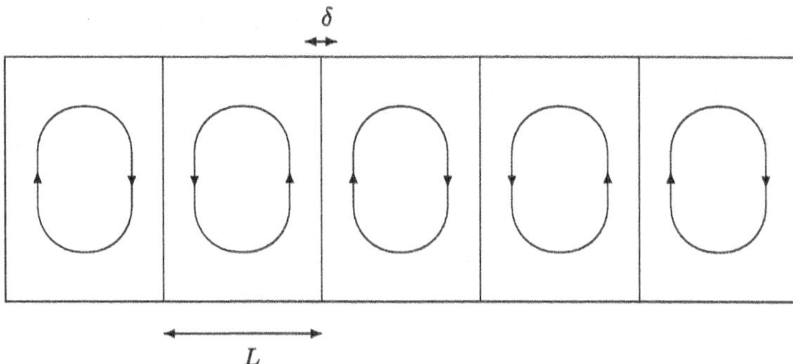

Fig. 7.7 Cellular flow. L is the size of the vortices, δ is the thickness of the diffusive boundary layer

differences, the fluid remains stationary and molecular diffusion transports heat from the lower to the upper plane. As the temperature difference increases, at a certain point an instability is observed (studied by Lord Rayleigh in 1916) and ordered convective motions arise in which regions of rising hot fluid alternate with regions of descending cold fluid, with the pattern shown in Fig. 7.7.

Apart from the physical motivations, the cellular flow has become a prototype in the study of transport in fluids because it presents *barriers* to transport given by the separatrices between the vortices. It is indeed evident that in the absence of molecular diffusion ($D = 0$), each trajectory remains trapped in its initial vortex and horizontal transport is zero. The situation changes drastically if we instead consider $D > 0$. Let us consider a typical rotation time of a trajectory, given by $\tau \sim L/u_0$ (with $u_0 = k\psi_0$ the typical velocity of the vortex). During this time, molecular diffusion will cause the trajectories to spread over a length $\delta \sim \sqrt{D\tau}$, so only the particles at a distance less than δ from the barrier will be able to change cell. These particles will then be transported by the velocity across the cell and will therefore have a "diffusion coefficient" of order L^2/τ. Since in an incompressible fluid the particle density is uniform, the fraction participating in the process will simply be L/δ and ultimately the effective diffusion coefficient will be

$$D^E \sim \frac{\delta}{L}\frac{L^2}{\tau} \sim \sqrt{DLu_0}$$

that is, it is proportional to the square root of the molecular diffusivity and is typically much greater than D. This result, which can be made rigorous by means of more advanced mathematics, has also been confirmed by a series of laboratory experiments and shows how the combined effect of the deterministic part (vortex velocity) and the random part (thermal noise) can lead to a nontrivial result.

7.5 Ergodicity, Statistical Mechanics, and Probability

We conclude this chapter with a discussion on the role of ergodicity in statistical mechanics and probability.

7.5.1 *Ergodic Hypothesis and Foundations of Statistical Mechanics*

Macroscopic systems are composed of a very large number (on the order of Avogadro's number) of particles; this fact makes it practically necessary to use a statistical description, in terms of "statistical ensembles," that is, by using probability distributions in phase space.

If $\mathbf{q}_i$ and $\mathbf{p}_i$ denote, respectively, the position and momentum vectors of the i-th particle, the state of a system of N particles is represented, at time t, by a vector $\mathbf{X}(t) \equiv (\mathbf{q}_1(t), \ldots, \mathbf{q}_N(t), \mathbf{p}_1(t), \ldots, \mathbf{p}_N(t))$ in a $6N$-dimensional space, called phase space. The observables of the system are represented by functions, $A(\mathbf{X})$, defined in phase space. Suppose that the Hamiltonian does not depend explicitly on time, then the energy is a conserved quantity during the motion, which therefore evolves on a hypersurface of fixed energy. Denoting by $V(\{\mathbf{q}_j\})$ the interaction potential between the particles, the Hamiltonian function is written as

$$H = \sum_{i=1}^{N} \frac{\mathbf{p}_i^2}{2m} + V(\{\mathbf{q}_j\}),$$

and the equations of motion are:

$$\frac{d\mathbf{q}_i}{dt} = \frac{\partial H}{\partial \mathbf{p}_i} = \frac{\mathbf{p}_i}{m}, \quad \frac{d\mathbf{p}_i}{dt} = -\frac{\partial H}{\partial \mathbf{q}_i} = -\frac{\partial V}{\partial \mathbf{q}_i} \tag{7.23}$$

with $i = 1, \ldots, N$.

It is fundamental to note that the timescale of macroscopic observations is much larger than the timescale of the microscopic dynamics (7.23), which dictates the rate of changes at the molecular level. This means that an experimental measurement is actually the result of a single observation during which the system passes through a very large number of different microscopic states. If the measurement refers to the observable $A(\mathbf{X})$, it should therefore be compared with an average taken along the evolution of the system and calculated over very long times (from the microscopic point of view):

$$\overline{A}(t_0, T) = \frac{1}{T} \int_{t_0}^{t_0+T} A(\mathbf{X}(t))\mathrm{d}t. \tag{7.24}$$

The calculation of the time average $\overline{A}$ requires, in principle, both knowledge of the complete microscopic state of the system at a certain instant and the determination of the corresponding trajectory in phase space. This requirement is clearly impossible; therefore, if $\overline{A}$ depended very strongly on the initial state of the system, it would not even be possible to make statistical predictions, even disregarding the difficulty of finding the solution to system (7.23). The ergodic hypothesis allows us to overcome this obstacle. It essentially states that every hypersurface of fixed energy is completely accessible to any motion with the given energy; that is, a constant-energy hypersurface cannot be divided into (measurable) regions each containing complete motions, i.e., invariant under time evolution (if this condition is satisfied, the hypersurface is said to be *metrically indecomposable* or *metrically transitive*). Furthermore, for every trajectory, the average time spent in a certain region is proportional to the volume of that region. If the above conditions, which constitute the core of the ergodic hypothesis, are satisfied, it follows that, if T is sufficiently large, the average in (7.24) depends only on the energy of the system and thus takes the same value for all evolutions with the same energy; moreover, this common value can be calculated by averaging $A(\mathbf{X})$ in which all (and only) the states with the fixed energy contribute with equal weight. The uniform probability density on the surface of fixed energy defines the microcanonical measure; denoting this density by $P_{mc}(\mathbf{X})$, the ergodic hypothesis allows us to write:

$$\overline{A} \equiv \lim_{T\to\infty} \frac{1}{T} \int_{t_0}^{t_0+T} A(\mathbf{X}(t))\mathrm{d}t = \int A(\mathbf{X}) P_{mc}(\mathbf{X})\mathrm{d}\mathbf{X} \equiv \langle A \rangle. \tag{7.25}$$

It should be emphasized that the validity of the previous equation simultaneously frees us from the need to determine an (initial) state of the system and to solve the equations of motion. The validity (or not) of (7.25), that is, the possibility of replacing the time average with a phase space average, constitutes the ergodic problem. Since an isolated system can be described by the microcanonical ensemble, from which, if $N \gg 1$, the canonical ensemble can be derived (see Chap. 2), the proof of (7.25) can be considered the dynamical justification for the introduction of statistical ensembles.

7.5.2 The Ergodic Problem and Analytical Mechanics

From what has been discussed above, regarding the metric indecomposability of the constant energy surface, the connection between the study of the validity of (7.25) and the search for the existence of constants of motion other than energy becomes clear. In fact, if there existed first integrals in addition to energy, the system would certainly not be ergodic, as it would not be metrically transitive: by choosing as observable A one of the first integrals, we would have $\overline{A} = A(\mathbf{X}(0))$, which depends on the initial condition $\mathbf{X}(0)$ and is therefore generally different from $\langle A \rangle$.

Given a Hamiltonian $H(\mathbf{q}, \mathbf{p})$, with $\mathbf{q}, \mathbf{p} \in \mathbb{R}^N$, if there exists a canonical transformation that takes the variables $(\mathbf{q}, \mathbf{p})$ to the action-angle variables $(\mathbf{I}, \boldsymbol{\phi})$, in such a way that the Hamiltonian depends only on the actions $\mathbf{I}$:

$$H = H_0(\mathbf{I}), \tag{7.26}$$

then the system is said to be integrable. In this case, the time evolution of the system is simple to write:

$$\begin{cases} I_i(t) = I_i(0) \\ \phi_i(t) = \phi_i(0) + \omega_i(\mathbf{I}(0))\, t, \end{cases}$$

where $\omega_i = \partial H_0 / \partial I_i$ and $i = 1, \ldots, N$. Note that in an integrable system there are N independent constants of motion, since all the actions I_i are conserved, and the motions take place on N-dimensional tori. An important example of an integrable system is provided by the Solar System, if the interactions between the planets are neglected: in this limit, the problem reduces to the two-body problem (Sun-planet), for which integrability is easily demonstrated.

It is natural to ask about the effect of a perturbation to (7.26), that is, to study the Hamiltonian

$$H(\mathbf{I}, \boldsymbol{\phi}) = H_0(\mathbf{I}) + \epsilon H_1(\mathbf{I}, \boldsymbol{\phi}). \tag{7.27}$$

In the case of the Solar System, this would mean taking into account the interactions between the planets, and we would have $\epsilon \sim 10^{-3}$, which is the ratio between the mass of Jupiter (the largest planet) and the Sun. Are the trajectories of the perturbed system (7.27) 'close' to those of the integrable system (7.26)? And does the introduction of the term $\epsilon H_1(\mathbf{I}, \boldsymbol{\phi})$ still allow for the existence of constants of motion in addition to energy?

These questions are of obvious interest for celestial mechanics, and also for the ergodic problem. Note that in statistical mechanics and in celestial mechanics, the desired behaviors are opposite: in statistical mechanics, one would like "irregular" dynamical behaviors, in order to justify the ergodic hypothesis; on the contrary, in celestial mechanics, "regular" behaviors are desirable, so that accurate predictions are possible.

In a very important work from 1890, Poincaré showed that, in general, a system of the type (7.27), with $\epsilon \neq 0$, does not admit analytic first integrals other than energy. Such a result seems very negative as far as celestial mechanics is concerned; however, we will see that this conclusion is by no means obvious. In 1923, Fermi generalized Poincaré's theorem, showing that in a generic Hamiltonian system with $N > 2$ there does not exist any $2N - 2$ dimensional surface invariant under time evolution, which is analytic in $\mathbf{I}$, $\boldsymbol{\phi}$ and ϵ.

From these results, Fermi argued that Hamiltonian systems in general (that is, apart from the integrable cases, which should be considered atypical) are ergodic as soon as $\epsilon \neq 0$; this conclusion (which later turned out to be incorrect) was essentially accepted by everyone (at least in the physics community).

7.5.3 An Unexpected Result

After the war, Fermi reconsidered the problem of ergodicity in the work *Studies of non linear problems*, the result of a collaboration with Pasta and Ulam.[4] Following common usage, we will refer to this work using the acronym FPU.

In the FPU, the evolution of a chain of $N + 2$ identical particles, of mass m, connected by nonlinear springs with Hamiltonian

$$H = \sum_{i=0}^{N} \left[\frac{p_i^2}{2m} + \frac{K}{2}\left(q_{i+1} - q_i\right)^2 + \frac{\epsilon}{r}\left(q_{i+1} - q_i\right)^r \right]$$

with $q_0 = q_{N+1} = p_0 = p_{N+1} = 0$ and $r = 3$ or $r = 4$. For $\epsilon = 0$ the system is integrable, as it is equivalent to N independent harmonic oscillators. In fact, by introducing the normal modes:

$$a_k = \sqrt{\frac{2}{N+1}} \sum_i q_i \sin \frac{i\,k\,\pi}{N+1} \qquad (k = 1, \dots, N),$$

one obtains N non-interacting harmonic oscillators, with angular frequencies

$$\omega_k = 2\sqrt{\frac{K}{m}} \sin \frac{k\,\pi}{2(N+1)}.$$

In this case, the energies of the normal modes

$$E_k = \frac{1}{2}\left(\dot{a}_k^2 + \omega_k^2 a_k^2\right).$$

are constants of motion, proportional to the action variables $E_k = \omega_k I_k$.

When considering small values of ϵ, it is not difficult to calculate all the thermodynamically relevant quantities within the framework of statistical mechanics, that is, by averaging over a statistical ensemble (for example, the microcanonical or

[4] The article, written as an internal note at the Los Alamos laboratories, was completed in May 1955, after Fermi's death, but appeared for the first time only in 1965, as contribution No. 266 to *Note e Memorie*, a collection of Fermi's writings published by the Accademia dei Lincei and the University of Chicago.

canonical ensemble). In particular, one has

$$\langle E_k \rangle \simeq \frac{E_{TOT}}{N}, \tag{7.28}$$

for $\epsilon = 0$ the equality is exact (equipartition of energy law). It should be noted that (7.28) is valid for $\epsilon = 0$ or small; however, in order for $\overline{E_k}$ (i.e., the average calculated along the trajectory) to coincide with $\langle E_k \rangle$, ϵ must necessarily be different from zero, so as to allow the normal modes to interact with each other, thus enabling the initial conditions to be 'forgotten'.

What happens if one chooses an initial configuration in which the energy is concentrated on a few normal modes, for example $E_1(0) \neq 0$ and $E_k(0) = 0$ for $k = 2, \ldots, N$? Before the FPU work, practically everyone would have expected (on the basis of the previous discussion on Poincaré's theorem and its extension due to Fermi) that the first normal mode would gradually transfer energy to the others and that, after a certain relaxation time, each $E_k(t)$ would fluctuate around the common mean value, given by (7.28). Fermi was probably interested in a numerical simulation, not so much to check his "proof" of the ergodic hypothesis, but rather to investigate the thermalization times, that is, the time needed for the system to go from a situation out of statistical equilibrium (all the energy concentrated in a single mode) to the equipartition situation, as expected from statistical mechanics. Contrary to expectations, in the FPU experiment no tendency toward equipartition was found, even after very long times. In other words, what was actually observed was a violation of ergodicity. The behavior, as a function of time, of the quantity E_k/E_{TOT}, for various k, does not show a loss of memory of the initial condition; on the contrary, after a certain (long) time, the system returns almost to its initial state.

The results obtained in the FPU experiment are decidedly contrary to what was expected, and Fermi himself, according to Ulam, was surprised and expressed the opinion that they were facing an important discovery that showed, unequivocally, how the prevailing beliefs about the generic nature of mixing and thermalization properties in nonlinear systems might not always be justified.

7.5.4 Theorems and Simulations

The "solution" to the problem, which constitutes an important chapter of modern mathematical physics, had (paradoxically) already been found a year before the drafting of the Los Alamos note (unbeknownst to the authors of the FPU), by Kolmogorov, who in 1954 stated (without presenting a detailed proof, but with the main idea clearly expressed) an important theorem, whose proof was later completed by Arnold and Moser. The theorem, now known by the acronym KAM, can be stated as follows:

Given a Hamiltonian $H(\mathbf{I}, \boldsymbol{\phi}) = H_0(\mathbf{I}) + \epsilon H_1(\mathbf{I}, \boldsymbol{\phi})$, with $H_0(\mathbf{I})$ sufficiently regular and also $\det|\partial^2 H_0(\mathbf{I})/\partial I_i \partial I_j| \neq 0$, if ϵ is small then on the constant energy surface there survive invariant tori (which are called KAM tori and are a small deformation of those present for $\epsilon = 0$) in a set whose measure tends to 1 as $\epsilon \to 0$.

The KAM theorem might seem intuitive, if one were not aware of the theorems on the non-existence of nontrivial first integrals. On the contrary, the existence of KAM tori is a very delicate and highly counterintuitive fact. Actually, for every value of ϵ (even small), some tori of the unperturbed system, those called resonant, are destroyed, and this prevents the existence of analytic integrals of motion. Nevertheless, when ϵ is small, the majority some of the tori, slightly deformed, survive; therefore, the perturbed system (for "non-pathological" initial conditions) behaves in a way not very different from the integrable case.

After the FPU, there was a long series of works (mainly numerical) that further investigated the problem as ϵ varies, in an attempt to reconcile the KAM theorem with the intuition of statistical mechanics. For a fixed number of particles N, the scenario is essentially as follows: at a given energy density $\mathcal{E} = E/N$, there exists a threshold ϵ_c for the intensity of the perturbation such that

(a) if $\epsilon < \epsilon_c$ the KAM tori are predominant and the system does not equipartition;
(b) if $\epsilon > \epsilon_c$ the KAM tori are negligible, the system does equipartition, and there is agreement with standard statistical mechanics.

It is easy to see that if, more in line with physical situations, the value of the perturbation ϵ is fixed, the energy density becomes the control parameter and there exists a threshold value $\mathcal{E}_c$ that separates regular from chaotic behavior.

Even after more than half a century since FPU, the understanding of the relevance of KAM, and more generally of chaos, for statistical mechanics still presents some unresolved aspects. In particular, from a physical point of view, several issues arise, for example:

(1) the dependence of ϵ_c on N (at fixed $\mathcal{E}$), or equivalently, what is the dependence of $\mathcal{E}_c$ on N (at fixed ϵ);
(2) what are the characteristic timescales of the equipartition process.

7.5.5 Is Ergodicity Really Necessary?

There is a school of thought (whose main exponents include Khinchin and Landau) that considers the whole issue of ergodicity to be essentially irrelevant in the context of statistical mechanics, since the ergodic hypothesis would in fact not be necessary to justify Eq. (7.25) for physically relevant observables. This point of view is based on the following facts:

(a) in systems of interest to statistical mechanics, the number of degrees of freedom is very large;
(b) the issue of interest for statistical mechanics is the validity of (7.25) not for any observable, but for the few quantities relevant to thermodynamics;
(c) it is physically acceptable to admit that ergodicity does not hold in a "small" region of phase space.

The conclusions obtained by Khinchin are summarized in the following result:[5]

If the observable A can be expressed as a sum of N components, each depending on the variables of a single particle

$$A = \sum_{i=1}^{N} f(\mathbf{q}_i, \mathbf{p}_i)$$

then for systems with Hamiltonian of the form

$$H = \sum_{n=1}^{N} H_n(\mathbf{q}_n, \mathbf{p}_n)$$

one has

$$P\left(\frac{|\overline{A} - \langle A \rangle|}{|\langle A \rangle|} \geq C_1 N^{-1/4}\right) \leq C_2 N^{-1/4},$$

where C_1 and C_2 are constants of order $O(1)$. Therefore, the relative measure (that is, the probability with respect to the microcanonical probability density P_{mc}) of the points, on the hypersurface of fixed energy, for which $\overline{A}$ is significantly different from $\langle A \rangle$ is a small quantity of order $N^{-1/4}$.

We can say that, in the limit $N \to \infty$, (7.25) is valid (except in a region of phase space that becomes ever smaller as N increases) for an interesting class of functions; and this is independent of the details of the dynamics.

To summarize: the results of Khinchin (Mazur and van der Linden) suggest that, for statistical mechanics, dynamics plays a marginal role, while the fundamental ingredient is that $N \gg 1$.

[5] Khinchin's result was originally proven for systems of non-interacting particles, and extended by Mazur and van der Linden to systems of interacting particles with short-range potentials.

7.6 Final Remarks on Chaos, Ergodicity, and Statistical Ensembles

Let us summarize the main conceptual aspects of this chapter.

Let us begin by emphasizing that the (re)discovery of deterministic chaos, i.e. the possibility of having a sensitive dependence on initial conditions, with aperiodic and apparently irregular behaviors, even in low-dimensional deterministic systems, has made it possible to reconsider some fundamental ideas about the real relevance of determinism and the statistical description of systems.

On a more specific level, we note that ergodicity, despite all the caveats we have mentioned, allows us to introduce probability in a very natural way in the context of deterministic systems. In particular, since ergodicity is equivalent to assuming that different trajectories have the same asymptotic properties, we have an unambiguous way to to interpret probability in frequentist terms. Assuming ergodicity, the probability of an event (defined through its asymptotic frequency) is an objective and measurable property, obtained from a trajectory.

We emphasize that we are talking about a *single* system (even if it has many degrees of freedom), and not about a collection of identical systems.

It seems natural to us, adopting Boltzmann's point of view, to consider that the only physically well-founded statistical approach (at the conceptual level) is the one in terms of averages obtained by following the time evolution of the system. Note that this is what is explicitly done in numerical simulations (both with the Monte Carlo method and with molecular dynamics) and also in experiments. In this way of looking at the problem, the introduction of statistical ensembles is not at all unnatural or mysterious,[6] a statistical ensemble is nothing more than a probability density in phase space, and is introduced only for practical reasons. Time averages, which are the only physically well-founded ones, are equivalent to averages with respect to a certain probability density in phase space, and this allows explicit calculation, or at least facilitates the task.

The microcanonical ensemble, in which the energy and the number of particles are fixed, has a special status: the microcanonical density is invariant with respect to the evolution given by Hamilton's equations and is closely linked to the ergodic hypothesis. The canonical ensemble, where the number of particles is fixed, is obtained by a marginalization procedure starting from the microcanonical, in this way one has energy fluctuations (percentage-wise small in the limit $N \gg 1$). Similarly, the grand canonical ensemble, in which the number of particles is not fixed, can be obtained by a marginalization procedure starting from the canonical, obtaining (asymptotically negligible) fluctuations both for the energy and for the number of particles.

---

[6] In some statistical mechanics textbooks one can find obscure statements such as *A statistical ensemble is a collection of identical systems*. It is not difficult to convince oneself, for example from Gibbs' book *Elementary Principles in Statistical Mechanics*, that *a collection of identical systems* is just a way to indicate the probability density.

There are situations in which even macroscopic systems are not ergodic (not even in a weak sense). These situations are very important (for example, glasses) and very difficult to deal with.

We conclude by saying that not everyone shares the above-described way of introducing statistical ensembles; see Chap. 9 for a brief discussion of a completely different point of view.

Exercises

7.1 Show that the map

$$x_{t+1} = x_t + \omega \ mod\ 1$$

is ergodic with respect to the uniform probability density in [0, 1] if ω is irrational, while it is not ergodic if ω is rational.

7.2 Show that the map

$$x_{t+1} = \begin{cases} x_t + 3/4 & if\ 0 \le x_t < 1/4 \\ x_t + 1/4 & if\ 1/4 \le x_t < 1/2 \\ x_t - 1/4 & if\ 1/2 \le x_t < 3/4 \\ x_t - 3/4 & if\ 3/4 \le x_t < 1 \end{cases}$$

is not ergodic with respect to the uniform probability density in [0, 1] which is invariant under the dynamics.

Hint: note that there exists an invariant set with respect to the dynamics.

7.3 Consider the map

$$x_{t+1} = \begin{cases} x_t/p & if\ 0 \le x_t < p \\ (1 - x_t)/(1 - p) & if\ p \le x_t < 1 \end{cases}$$

(a) show that the invariant probability density is constant in [0, 1];
(b) numerically study the time evolution of the probability density starting from

$$\rho_o(x) = \begin{cases} 1/\Delta & if\ \ x \in [x_0, x_0 + \Delta] \\ 0 & otherwise \end{cases}$$

Consider various choices of x_0 and Δ and compare the results with the limiting density $\rho_{inv}(x) = 1$.

7.4 Consider the map

$$x_{t+1} = \begin{cases} 2x_t & if\ 0 \le x_t < 1/2 \\ x_t - 1/2 & if\ 1/2 \le x_t < 1 \end{cases}$$

and the Markov partition $A_0 = [0, 1/2]$ and $A_1 = [1/2, 1]$. Calculate the invariant probability density using the Perron-Frobenius operator and check that you get the same result from the Markov chain associated with the partition (A_0, A_1).

7.5 Numerically study the time evolution of the probability density of the logistic map

$$x_{t+1} = 4x_t(1 - x_t)$$

starting from the initial density

$$\rho_o(x) = \begin{cases} 1/\Delta & if \;\; x \in [x_0, x_0 + \Delta] \\ 0 & \text{otherwise} \end{cases}$$

Consider various choices of x_0 and Δ and compare the results with the limiting density $\rho_{inv}(x) = 1/\pi\sqrt{x(1-x)}$.

Recommended Readings

For a simple introduction to deterministic chaos:

A. Lasota, M.C. Mackey, *Probabilistic Properties of Deterministic Systems*, (Cambridge University Press, Cambridge, 1985)

For an introduction to diffusion:

G. Boffetta, G. Lacorata, A. Vulpiani, Introduction to chaos and diffusion, in *Chaos in Geophysical Flows*, ed. by G. Boffetta, G. Lacorata, G. Visconti, A. Vulpiani (Otto Eds. Torino, 2003), p. 5. http://arxiv.org/abs/nlin/0411023

For a general discussion on chaos and ergodicity in statistical mechanics:

P. Castiglione, M. Falcioni, A. Lesne, A. Vulpiani, *Chaos and Coarse Graining in Statistical Mechanics* (Cambridge University Press, Cambridge, 2008)

Khinchin's results are in the book already cited in Chap. 2

A nice book that deals with many conceptual aspects in statistical mechanics, probability, and chaos:

G.G. Emch, C. Liu, *The Logic of Thermostatistical Physics* (Springer, Berlin, 2001)

A collective volume on the FPU:

G. Gallavotti (ed.), *The Fermi-Pasta-Ulam Problem.* Springer Lecture Notes in Physics, vol. 728 (2008); in particular the detailed contribution A.J. Lichtenberg, R. Livi, M. Pettini, Ruffo, "Dynamics of oscillator chains", p. 21

For glassy systems:

M. Mezard, G. Parisi, M. Virasoro, *Spin Glass Theory and Beyond* (World Scientific, Singapore, 1987)

Chapter 8
Beyond the Gaussian Distribution

In Chap. 3 we discussed in detail how the Central Limit Theorem (CLT) holds under very general assumptions, and therefore the Gaussian probability distribution (and distributions closely related to it, such as the lognormal) appears in many contexts (for example, in many problems in physics, chemistry, geology). However, this should not lead the reader to make undue extrapolations, concluding that there are no other probability distributions that appear non-artificially in problems of physical interest. In the first part of this chapter, we will discuss an important generalization of the CLT for variables with infinite variance. In the second part, we will briefly address some physical problems in which non-Gaussian probability distributions naturally appear.

8.1 Some Observations

In Chap. 3, when discussing the sum of random variables, we saw, with the help of the generating function and the characteristic function, an interesting property: the sum of Gaussian variables is still a Gaussian variable. Similarly, the sum of Poisson variables is still a Poisson variable.

Indicating with $p_{m,\sigma^2}(x)$ the probability density of a Gaussian variable with mean m and variance σ^2, we have for the convolution

$$\left(p_{m_1,\sigma_1^2} \star p_{m_2,\sigma_2^2}\right)(x) = p_{m,\sigma^2}(x) \tag{8.1}$$

with $m = m_1 + m_2$ and $\sigma^2 = \sigma_1^2 + \sigma_2^2$. This type of property also holds for variables with the Pearson distribution (see the Appendix for definitions)

$$p_{\lambda,N}(x) = \frac{\lambda^N}{(N-1)!} x^{N-1} e^{-\lambda x}$$

G. Boffetta, A. Vulpiani, *Probability in Physics*, UNITEXT for Physics,
https://doi.org/10.1007/978-3-032-10407-6_8

it is in fact immediate to verify that

$$\left(p_{\lambda,N_1} \star p_{\lambda,N_2}\right)(x) = p_{\lambda,N_1+N_2}(x). \tag{8.2}$$

In statistical mechanics, (8.2) has a well-defined physical meaning. In a perfect gas at thermal equilibrium at temperature T, containing N particles, the probability density of the energy x is

$$p_n(x) = \frac{\lambda^n}{(n-1)!} x^{n-1} e^{-\lambda x}$$

where $n = 3N/2$ and $\lambda = \beta$. Equation (8.2) corresponds to the fact that, by bringing into contact two systems with N_1 and N_2 particles, both at thermal equilibrium at the same temperature $T = 1/\beta$, one obtains a system of $N = N_1 + N_2$ particles in thermodynamic equilibrium at the same temperature $T = 1/\beta$.

Similarly, for the Cauchy distribution:

$$p_{m,a}(x) = \frac{a^2}{\pi[(x-m)^2 + a^2]}$$

we have

$$\left(p_{m_1,a_1} \star p_{m_2,a_2}\right)(x) = p_{m_1+m_2,a_1+a_2}(x), \tag{8.3}$$

and also for Poisson variables with

$$P_\lambda(k) = \frac{\lambda^k}{k!} e^{-k}$$

it is easy to show that

$$\left(P_{\lambda_1} \star P_{\lambda_2}\right)(k) = P_{\lambda_1+\lambda_2}(k). \tag{8.4}$$

The validity of (8.1), (8.2), (8.3), and (8.4) is a consequence of the particular form of the corresponding characteristic functions. For the Gaussian, Cauchy, Pearson, and Poisson distributions, we have respectively

$$\phi_G(t) = e^{imt - \frac{\sigma^2}{2} t}, \quad \phi_C(t) = e^{imt - a|t|},$$

$$\phi_{Pe}(t) = \frac{1}{(1 - it/\lambda)^N}, \quad \phi_{Po}(t) = e^{\lambda(e^{it}-1)}.$$

Recalling that the characteristic function of a sum of random variables is given by the product of the characteristic functions, it is immediate to verify (8.1), (8.2), (8.3), and (8.4).

We conclude these initial observations by noting that the Binomial probability

$$P_{p,N}(k) = \frac{N!}{(N-k)!k!}\, p^k(1-p)^{N-k} \quad k = 0, 1, \ldots, N$$

also enjoys the property

$$\left(P_{p,N_1} \star P_{p,N_2}\right)(k) = P_{p,N_1+N_2}(k).$$

From this relation, recalling that the Gaussian and Poisson variables are limiting cases of the Binomial, in an analogous way the Pearson distribution is closely connected with the Poisson distribution (see Chap. 10), so, a posteriori, the validity of relations (8.1), (8.2), and (8.4) is not surprising.

8.2 Infinitely Divisible Probability Distributions and Stable Distributions

Let us observe that the previous examples of probability distributions that remain invariant under the operation of convolution (with appropriate "rescaling of the parameters") all have, except for the Cauchy distribution, finite variance.

From the Central Limit Theorem, we know that among probability distributions with finite variance, the Gaussian enjoys a special status: it is the only one that is "attractive." In fact, if we consider N i.i.d. variables with mean m and finite variance σ, for $N \gg 1$ the variable

$$z = \frac{1}{\sigma\sqrt{N}} \sum_{j=1}^{N} (x_j - m) \tag{8.5}$$

has a Gaussian probability density with zero mean and unit variance.

Let us now try to go beyond the observations made.

A variable Y is said to be infinitely divisible if for every N it can be written as the sum of i.i.d. variables:

$$Y = X_1 + X_2 + \ldots + X_N.$$

In other words, Y is infinitely divisible if the convolution of its probability density is invariant (apart from a rescaling of the parameters). It is easy to see that its characteristic function $\phi_Y(t)$ must be the N-th power of a characteristic function $\phi_X(t, \frac{1}{N})$

$$\phi_Y(t) = \left[\phi_X\left(t, \frac{1}{N}\right)\right]^N.$$

For example, for Cauchy variables with $a = 1$ and $m = 0$ we have

$$\phi_Y(t) = e^{-|t|} = \left[e^{-\frac{1}{N}|t|}\right]^N.$$

It has been shown that the most general form of the characteristic function of an infinitely divisible distribution is given by

$$\ln \phi(t) = it\alpha + \int_{-\infty}^{\infty} \left(e^{itx} - 1 - \frac{itx}{1+x^2}\right) \frac{1+x^2}{x^2} dG(x) \tag{8.6}$$

where α is a constant and $G(x)$ is a real, bounded, non-decreasing function with $G(-\infty) = 0$. Equation (8.6) is called the Lévy-Khinchin formula. Noting that

$$\lim_{x \to 0} \left(e^{itx} - 1 - \frac{itx}{1+x^2}\right) \frac{1+x^2}{x^2} = -\frac{t^2}{2},$$

it is easy to verify that by taking $\alpha = m$ and $G(x) = \sigma\Theta(x)$ (where Θ is the step function) we obtain the Gaussian, and similarly with $\alpha = m$ and $G(x) = a/2 + (a/\pi)\tan^{-1}(x)$ we obtain the Cauchy distribution.

We are now ready for the generalization of the Central Limit Theorem for variables with infinite variance. The argument, which we outline without any claim of rigor, is entirely analogous to the one seen in Chap. 3. Let us consider N non-negative i.i.d. variables $X_1, \ldots, X_N$ with a density with power-law tails, that is, for large $|x|$

$$p(x) \sim \frac{1}{|x|^{1+\mu}}$$

with $0 < \mu < 2$, so $E(x^2) = \infty$. If $1 < \mu < 2$, then $m = E(x) < \infty$ and for the Laplace transform of $p(x)$ the following approximation holds (see the Appendix at the end of this chapter)

$$\mathcal{L}[p](s) = \int_0^{\infty} p(x)e^{-sx}dx \simeq 1 - As^{\mu} - ms + \ldots \tag{8.7}$$

where the constant A depends on $p(x)$. Let us consider the zero-mean variable $x' = x - m$; to calculate its Laplace transform, it is enough to subtract the term ms:

$$\mathcal{L}_{x'}(s) \simeq 1 - As^{\mu} + \ldots$$

Now let us introduce the variable

$$y'_N = \frac{1}{(AN)^{\frac{1}{\mu}}} \sum_{j=1}^{N} (x_j - m)$$

which generalizes (8.5) to the case with infinite variance. Recalling (see Appendix) that the Laplace transform is a moment generating function, since the variables $\{X_j - m\}$ are independent, we have:

$$\mathcal{L}_{y'}(s) = \left(\mathcal{L}_{x'}(s')\right)^N, \quad s' = \frac{s}{(AN)^{\frac{1}{\mu}}}$$

therefore

$$\mathcal{L}_{y'}(s) \simeq \left(1 - \frac{s^\mu}{N} + \ldots\right)^N \simeq e^{-s^\mu}.$$

In the case $0 < \mu < 1$ Eq. (8.7) is always valid, see Appendix, where m is replaced by a positive constant, in this case the term As^μ is dominant with respect to the $O(s)$ term and therefore for the Laplace transform of the variable

$$y_N = \frac{1}{(AN)^{\frac{1}{\mu}}} \sum_{j=1}^{N} X_j$$

we have

$$\mathcal{L}_{y_N}(s) \simeq \left(1 - \frac{s^\mu}{N} + \ldots\right)^N \simeq e^{-s^\mu}.$$

We thus have a result very similar to the one seen in Chap. 3, but now the normalization factor (which in the case of variables with finite variance is $1/\sqrt{N}$) depends on $p(x)$, in particular on the power-law behavior for $|x| \gg 1$ and is $N^{-\frac{1}{\mu}}$.

At this point, invoking the equivalence between the Laplace transform and the original function, we can infer that the distribution of the variable Y_N for $N \gg 1$ is a function that does not depend on the shape of $p(x)$ but only on the value of μ (and on other parameters, such as the mean value).

The result just discussed heuristically has been proven rigorously. The limiting distribution, denoted by $L_\mu(z)$, is called the μ-stable Levy function. Its explicit form is not known (except for some values of μ), but its characteristic function is known:

$$\ln \phi(t) = imt - a|t|^\mu \left(1 - i\beta \frac{t}{|t|} \omega(t, \mu)\right)$$

where $\omega(t, \mu) = \tan(\pi\mu/2)$ if $\mu \neq 1$ while $\omega(t, \mu) = -\frac{2}{\pi} \ln |t|$ if $\mu = 1$. The parameter $a > 0$ is a "scale", m is the mean value, $\beta \in (-1, 1)$ is the asymmetry factor, and $\mu \in (0, 2]$. For $\mu = 2$ we have the Gaussian distribution, the Cauchy distribution corresponds to the case $\mu = 1$, and in general for $|z| \gg 1$ we have

$$L_\mu(z) \sim \frac{1}{|z|^{1+\mu}}.$$

We thus have a generalization of the CLT for independent variables with infinite variance: let $x_1, \dots, x_N$ be i.i.d. with tails of the type

$$p(x) \sim \frac{1}{|x|^{1+\mu}} \quad 0 < \mu < 2, \tag{8.8}$$

then the variable

$$y_N = \frac{1}{N^{\frac{1}{\mu}}} \sum_{j=1}^{N} (x_j - m)$$

if $N \gg 1$, is distributed as a Levy function L_μ where the parameters m and a depend on the details of $p(x)$. Note that the L_μ are not only infinitely divisible but also "attractive" (hence the name stable), in fact the L_μ are the limiting distributions of the repeated convolution (with appropriate rescaling $N^{-1/\mu}$) of probability densities with asymptotic behavior (8.8).

8.2.1 *An Example from Physics*

Let us briefly mention an example of a physical process in which Levy distributions appear. The well-known Arrhenius law states that the mean exit time τ from an energy barrier E is (see Chap. 6)

$$\tau(E) \sim \tau_0 e^{E/k_B T}$$

where τ_0 is a typical time of the process. In disordered systems, the barrier energy E is a random variable with probability density $p_E(E)$, and therefore $p_\tau(\tau)$ is determined by the relation

$$p_\tau(\tau) = \frac{1}{|\frac{d\tau(E)}{dE}|} p_E(E). \tag{8.9}$$

In many situations, $p_E(E)$ is an exponential function:

$$p_E(E) = \frac{1}{E_0} e^{-E/E_0} \tag{8.10}$$

and therefore, using (8.9) and (8.10) we have

$$p_\tau(\tau) \sim \frac{A}{\tau^{1+\mu}}$$

where $\mu = k_B T / E_0$. Note that if $\mu > 2$ then both $< \tau >$ and the variance are finite. More interesting is the case $\mu < 2$ (i.e., $k_B T < 2E_0$) in which the variance is infinite, while if $\mu < 1$ even $< \tau >$ is infinite. The variable

$$T_N = \sum_{n=1}^{N} \tau_n$$

is the time the system takes to make N transitions (total trapping time), of interest in transport problems in disordered systems.

8.3 Diffusion Processes Do Not Always Have a Gaussian Distribution

Levy distributions, although interesting from a mathematical point of view, do not have an immediate applicability in physics. The reason is obviously due to the infinite variance, which, although not physically impossible, as in the case of the waiting time in the previous example, is generally not compatible with most physical quantities.

Nonetheless, there are many problems of great practical interest for which the CLT may not hold even in the presence of distributions with finite variance finite. This occurs when we are in the presence of very long *temporal correlations* $C(\tau)$, that is, when $\int_0^\infty C(\tau) d\tau = \infty$. In these cases, also called *Levy walks*, the random variable results from the sum of non-independent random variables, so, as seen in Chap. 3, the Central Limit Theorem is not applicable.

In this section, we will discuss two examples of this "anomalous" behavior that appear in problems of tracer transport in a fluid. The first is the case of dispersion in turbulence, while the second is given by transport in a particular laminar flow.

8.3.1 Relative Dispersion in Turbulence

Turbulence in fluids is a problem of great practical interest, present in phenomena ranging from the flow around a moving body (car, ship, airplane) to planetary atmospheric and oceanic scales, and also in convective phenomena in stars. Among the fundamental properties of turbulence, here we focus on the well-known efficiency in mixing and transporting substances. It is well known that a very efficient way to mix milk with coffee (or vermouth with gin, depending on your preferences) is

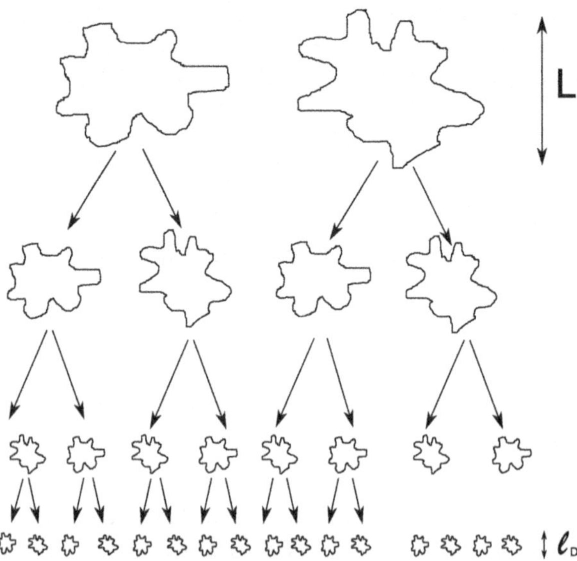

Fig. 8.1 Schematic representation of the turbulent cascade: the vortices created at the integral scale L by the external forcing, through hydrodynamic instabilities generate vortices at progressively smaller scales down to the dissipative scale ℓ_D

to stir with a spoon, which induces turbulent motions in the fluid.[1] These motions cause mixing down to the smallest scales, with times that are orders of magnitude shorter than the mixing due to molecular motion (Figs. 8.1 and 8.2).

The physical reason for this efficiency is that turbulence features motions at all scales generated by what is called the *turbulent cascade*. Without going into detail here (for which we refer the interested reader to the vast existing literature), let us recall that in the turbulent cascade there are statistical laws, originally due to Kolmogorov, for the fluctuations of velocity at a given scale. Let $\delta v(\ell)$ be the typical turbulent velocity fluctuation at a scale ℓ, that is, $\delta v(\ell) = v(x+\ell) - v(x)$ (where v represents a component of the velocity field $\mathbf{u}$), then Kolmogorov scaling holds:

$$\delta v(\ell) \sim \varepsilon^{1/3} \ell^{1/3} \tag{8.11}$$

where ε represents the energy density flowing through the cascade per unit time (energy flux). Note that (8.11) should be understood as a scaling law, in the sense that dimensionless prefactors are missing and it holds only in a statistical sense, that is, $\langle(\delta v(\ell_1))^2\rangle / \langle(\delta v(\ell_2))^2\rangle = (\ell_1/\ell_2)^{2/3}$. Kolmogorov scaling has been verified in countless experimental flows, most often by measuring the *energy spectrum* $E(k)$

[1] In reality, even non-turbulent velocity fields can produce very effective mixing, as in the example discussed at the end of this section.

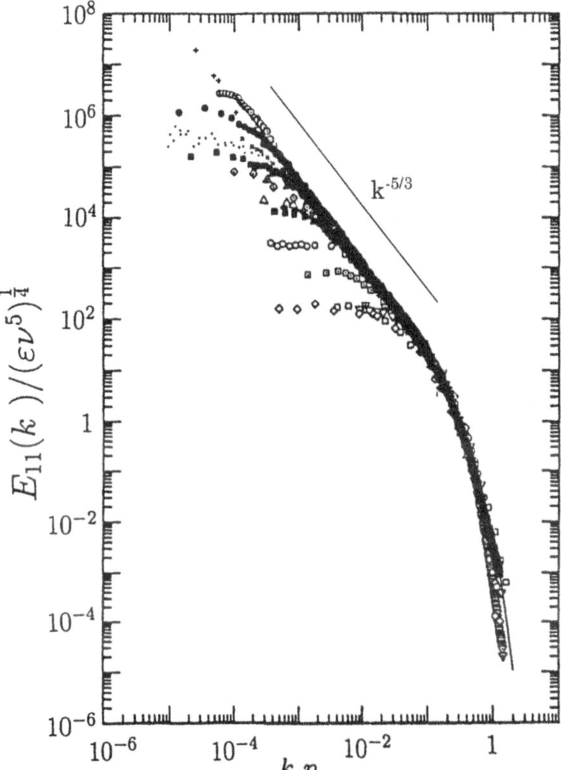

Fig. 8.2 Energy spectrum $E(k)$ for a series of turbulent flows at different Reynolds numbers as a function of the wavenumber k rescaled with the Kolmogorov scale $\eta = \ell_D$. The intermediate region clearly shows an inertial range with a power law $E(k) \sim k^{-5/3}$ as predicted by (8.12). As the Reynolds number increases, the width of the inertial range grows in accordance with $L/\ell_D \sim Re^{3/4}$. Figure taken from S.G. Saddoughi, S.V. Veeravall, *J. Fluid Mech.* **268**, 333 (1994)

(Fourier transform of the velocity correlator, where k represents the wavenumber, i.e., an inverse scale). The scaling (8.11) translates into the prediction of the Kolmogorov spectrum

$$E(k) = C\varepsilon^{2/3}k^{-5/3} \tag{8.12}$$

where C is a dimensionless constant to be determined experimentally.

The scaling law (8.11), or equivalently (8.12), holds between a maximum scale L, which is the velocity correlation scale and is determined by the geometry of the flow and the forcing that induces turbulence, and a minimum scale ℓ_D called the dissipative (or Kolmogorov) scale, at which turbulent velocity fluctuations are converted into heat by molecular viscosity. The range of scales $\ell_D < \ell < L$ of the cascade is called the *inertial range*, and its extent is determined by the Reynolds number Re as $L/\ell_D \sim Re^{3/4}$. Note that Re is typically very large for a macroscopic flow, for example $Re = O(10^6 - 10^7)$ for the flow around an automobile and $Re = O(10^9 - 10^{10})$ for large-scale atmospheric motions.

Let us now consider the *Lagrangian* problem of a small tracer particle (for example, a droplet of pollutant or an aerosol particle in the atmosphere) transported by a turbulent flow. As discussed in Chap. 7, the position of the particle is given by (7.19), which we rewrite as

$$\frac{d\mathbf{x}}{dt} = \mathbf{u}(\mathbf{x}, t) + \sqrt{2D}\eta. \tag{8.13}$$

Since the effect of the turbulent field is dominant compared to that of molecular diffusion, the last term in (8.13) can be neglected. There are two fundamental statistical aspects of Lagrangian dispersion: the *absolute dispersion*, that is, the mean square displacement of the particle with respect to the initial point discussed in Chap. 7; and the *relative dispersion*, given by the mean square separation between initially nearby particles. Relative dispersion describes the growth of the volume occupied by a set of particles due to turbulent transport and is fundamental in the study of pollutant dispersion. It is interesting to recall that the quantitative law for relative dispersion in turbulence was empirically determined by L.F. Richardson in 1926, a full 15 years before Kolmogorov's theory of turbulence (Fig. 8.3).

For the study of relative dispersion, we introduce the distance between two particles $\mathbf{r}(t) = \mathbf{x}_2(t) - \mathbf{x}_1(t)$. Following Richardson, we write for the probability density of separation $p(\mathbf{r}, t)$ the Fokker-Planck equation (6.16), which in the isotropic case becomes

$$\frac{\partial p(\mathbf{r}, t)}{\partial t} = \frac{1}{r^2}\frac{\partial}{\partial r}\left[r^2 K(r)\frac{\partial p(\mathbf{r}, t)}{\partial r}\right]$$

and we note that it must satisfy the normalization condition $\int p(\mathbf{r}, t) d\mathbf{r} = 1$. The diffusion coefficient in (8.14), $K(r)$, is given dimensionally, using (8.11), as $K(r) \simeq r\delta v(r) = k_0\varepsilon^{1/3}r^{4/3}$ (the dependence $K(r) \sim r^{4/3}$ was empirically obtained by Richardson from atmospheric data, and k_0 represents a dimensionless constant to be determined experimentally). The solution of (8.14) for an initial

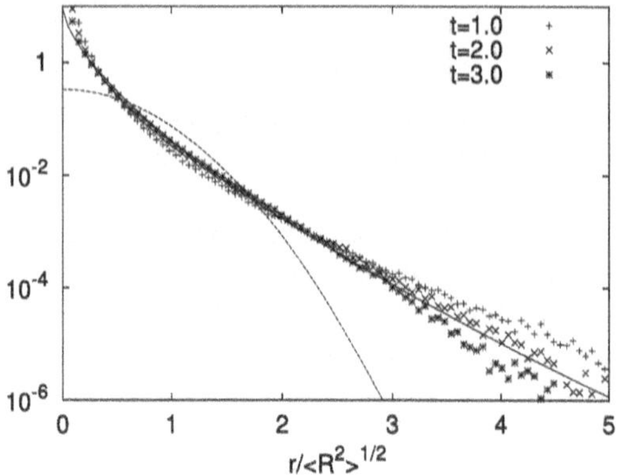

Fig. 8.3 Probability density of separation (in one component) at different times for pairs of particles transported by a turbulent flow obtained from a numerical simulation of the equations of motion for a fluid. The solid line is Richardson's prediction (8.14), the dashed line represents a Gaussian

distribution $p(\mathbf{r}, 0) = \delta(\mathbf{r})$ thus becomes

$$p(\mathbf{r}, t) = \frac{A}{(k_0 t)^{9/2} \varepsilon^{3/2}} \exp\left(-\frac{9r^{2/3}}{4k_0 \varepsilon^{1/3} t}\right) \tag{8.14}$$

where $A = 2187/(2240\pi^{3/2})$ is the normalization factor (see Fig. 8.3).

Therefore, we have that the distribution of separations in turbulence follows a non-Gaussian law and with a variance that grows faster than quadratically, that is, according to the law (called "Richardson's law")

$$\langle r^2(t) \rangle = \frac{1144}{81} k_0^3 \varepsilon t^3. \tag{8.15}$$

An intuitive explanation of the physical reason why the CLT does not apply in this case is that the evolution of the variable $\mathbf{r}(t)$ is given by a superposition of velocities that increase over time, since, according to (8.11) and (8.15), we have $\langle \delta v^2 \rangle \sim \varepsilon^{4/3} t^2$ and therefore we will have infinite variance for asymptotic times. In reality, in any real flow, the velocity variance is finite, since, as explained, the inertial range in which Kolmogorov scaling holds is limited above by the scale L. The fundamental reason why, for separations within the inertial range $\ell_D < r < L$, a standard diffusive process is not observed is that the requirement of temporal decorrelation at the basis of the CLT is not satisfied, as discussed in Chap. 3. In fact, according to Kolmogorov's theory, a velocity fluctuation at scale ℓ has a characteristic correlation time that scales dimensionally as $\tau(\ell) \sim \ell/\delta v(\ell) \sim \ell^{2/3}$, which therefore increases as the separation r between the particles increases.

8.3.2 Anomalous Diffusion in the Presence of Long Temporal Correlations

We now describe an example of anomalous diffusion obtained from the transport of particles in a random flow. The model was originally motivated by the study of fluid transport in porous media (for example, in aquifers) but has become a prototype of anomalous diffusion as a result of the combined effect of molecular diffusion and transport due to a velocity field.

For simplicity, let us consider a two-dimensional porous medium consisting of layers of constant thickness δ along the y direction, as shown in Fig. 8.4. Each layer is characterized by a velocity in the direction along the layer (i.e., along x) that is constant in time. We therefore have a horizontal velocity field that depends on the vertical coordinate, i.e., a random shear $u(y)$ which we will assume has zero mean. Superimposed on this field we have molecular diffusion (with diffusion coefficient D) and therefore, in accordance with (7.19), the motion of a particle is given by

$$\frac{dx}{dt} = u(y(t)) \qquad \frac{dy}{dt} = 2D\eta(t) \tag{8.16}$$

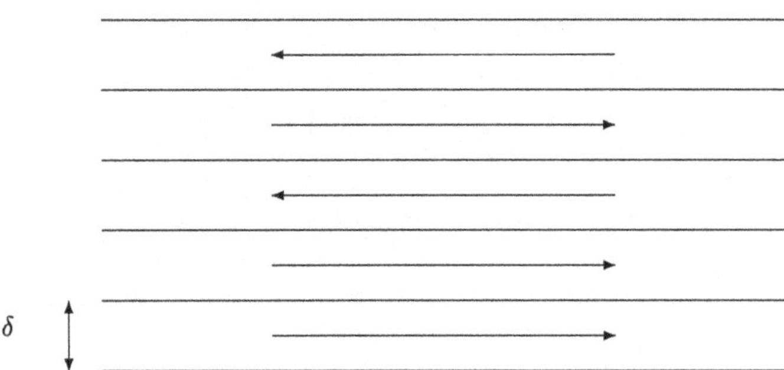

Fig. 8.4 Schematic representation of the random shear $u(y)$ with layers of thickness δ

where we have neglected the contribution of diffusion in the equation for x since in this direction the motion is dominated by the velocity field $u(y)$. For the stochastic term $\eta(t)$ we have (see Chap. 6) temporal correlation $\langle\eta(t)\eta(t')\rangle = \delta(t-t')$ and for simplicity we will assume that the velocity field $u(y)$ has a Gaussian distribution with zero spatial correlation, that is,

$$\langle u(y)u(y')\rangle_c = \sigma\delta(y-y') \tag{8.17}$$

where the notation $\langle\ldots\rangle_c$ indicates an average over all possible configurations of the medium.

Formally integrating (8.16) as $x(t) = \int_0^t u(y(t'))dt'$ and assuming $x(0) = y(0) = 0$, we have for the variance

$$\langle\langle x^2(t)\rangle_\xi\rangle_c = 2\int_0^t dt_2 \int_0^{t_2} dt_1 \langle\langle u(y(t_2))u(y(t_1))\rangle_\xi\rangle_c,$$

where $\langle\ldots\rangle_\xi$ denotes the average over different realizations of the molecular noise. The velocity correlation can be calculated by means of the probability that the random walk starting from $y = 0$ visits the points y_1 and y_2 at times t_1 and t_2, i.e., in terms of the transition probability defined in Chap. 6 as

$$\langle\langle u(y(t_2))u(y(t_1))\rangle_\xi\rangle_c =$$
$$\int_{-\infty}^{+\infty} dy_1 \int_{-\infty}^{+\infty} dy_2 P(0,0|y_1,t_1)P(y_1,t_1|y_2,t_2) \times \langle u(y(t_1))u(y(t_2))\rangle_c.$$

Since in the y direction we have a diffusive process, we have $P(y_0,t_0|y,t) = \frac{1}{\sqrt{4\pi D(t-t_0)}}e^{-(y-y_0)^2/(4D(t-t_0))}$, therefore, using (8.17) and performing the integrals, we obtain

$$\langle\langle x^2(t)\rangle_\xi\rangle_c = 2\int_0^t dt_2 \int_0^{t_2} dt_1 \frac{\sigma}{\sqrt{4\pi D(t_2-t_1)}} = \frac{4\sigma}{3\sqrt{\pi D}}t^{3/2} \tag{8.18}$$

that is, anomalous *superdiffusive* diffusion (i.e., with a temporal exponent for the variance greater than one). We observe that in the limit $D \to 0$ the variance diverges. In fact, if $D = 0$ the particle remains at $y = 0$ and therefore the motion is ballistic with $\langle x^2(t) \rangle \sim t^2$.

It can also be shown that the probability density has a non-Gaussian behavior

$$p(x,t) \sim \frac{1}{t^{3/4}} e^{-c\frac{x^4}{t^3}}$$

The origin of the anomalous behavior (8.18) can also be understood within the general framework of transport described in Chap. 7. The key point is that the velocity correlation function $C(\tau) = \langle u(\tau)u(0) \rangle$ in this case decays slowly as $C(\tau) \sim 1/\sqrt{\tau}$ (as seen in (8.18)), and therefore the integral of the Lagrangian autocorrelation does not converge and from (7.21) we have $\langle x^2(t) \rangle \sim t^{3/2}$. Note that the long temporal correlation of the velocity is due to the quenched nature of the realization of the velocity $u(y)$, which causes the velocity to "re-correlate" when the trajectory passes again through a given layer.

8.3.3 Anomalous Diffusion in a Vortex Array

Let us again consider the vortex chain model introduced in Chap. 7. Recall that this model was introduced to represent the velocity field generated by convective motions between two horizontal planes at different temperatures. As the temperature difference increases, it is observed experimentally that the stationary flow of Fig. 7.7 becomes unstable and the vortex chain begins to oscillate rigidly in the horizontal plane.

A good modeling of the flow in this regime is given by the stream function

$$\psi(x, y, t) = \psi_0 \sin(x + B \sin \omega t) \sin y \tag{8.19}$$

in which the parameter B represents the amplitude of the lateral oscillation of the vortices. Obviously, for $B = 0$ we recover the case studied in Chap. 7, so we now consider $B > 0$. Once B is fixed, the second control parameter of the system is the frequency of the oscillations, which can be made dimensionless by introducing the parameter $\epsilon = \omega L^2/\psi_0$ ($L = 2\pi$ is the width of a single cell). The two-dimensional velocity field associated with (8.19) will be given by $\mathbf{u}(\mathbf{x}, t) = (\partial_y \psi, -\partial_x \psi)$ and the trajectories $\mathbf{x}(t)$ of the transported particles will be given by the stochastic differential equation (7.19)

$$d\mathbf{x}(t) = \mathbf{u}(\mathbf{x}(t), t)dt + \sqrt{2D}\, d\mathbf{W}(t) \tag{8.20}$$

where D represents the molecular diffusion coefficient.

For $\epsilon > 0$, the system (8.19) and (8.20) can generate chaotic trajectories in which the particles can change cell even for $D = 0$ (unlike the stationary case). In general, therefore, at long times the horizontal motion of the particles is diffusive, with an effective (or turbulent) diffusion coefficient D^E that can be much greater than D.

The interesting aspect of this model is that for $\epsilon \simeq 1$ we can have a regime of synchronization between the frequency of lateral oscillation (ω) and the characteristic frequency of rotation in a vortex (of order ψ_0/L^2). Under these resonance conditions, the particle can "jump" to the adjacent cell at each period, and therefore the motion remains correlated for very long times, leading to an effective superdiffusive motion. This synchronization mechanism, similar to that of stochastic resonance discussed in Chap. 6, leads to an effective diffusion that is strongly dependent on the value of the parameter ϵ. In Fig. 8.5 we show the diffusion coefficient D^E_{11} in the x direction as a function of ϵ, for different values of the molecular coefficient D. We observe the presence of peaks, typical of synchronization conditions, whose height increases as D decreases. In the limit $D \to 0$, at the peaks, a superdiffusive regime appears for which $\langle x^2(t)\rangle \sim t^{2\nu}$ with $\nu > 1/2$, so that formally $D^E \to \infty$.

An example of this anomalous behavior is shown in Fig. 8.6, where the variance of the displacement along the x direction is shown as a function of time for a value $\epsilon = 1.1$ corresponding to of one of the peaks in Fig. 8.5. The dashed line represents

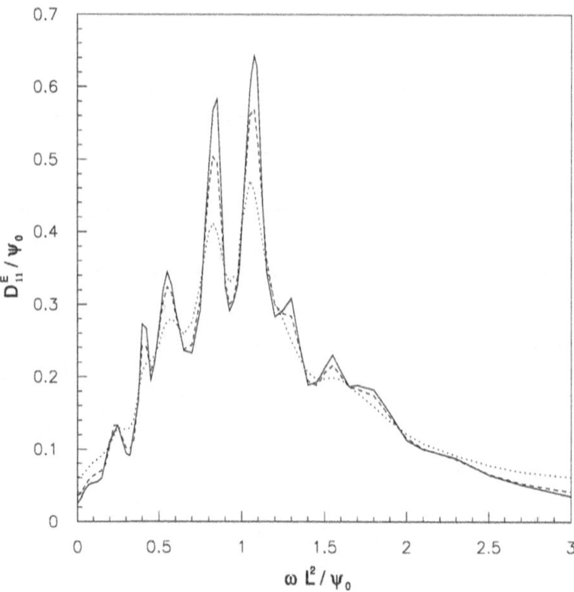

Fig. 8.5 Effective diffusion coefficient D^E as a function of the parameter $\epsilon = \omega L^2/\psi_0$ for different values of the diffusivity D. $D/\psi_0 = 3 \times 10^{-3}$ (dotted line), $D/\psi_0 = 10^{-3}$ (dashed line), and $D/\psi_0 = 5 \times 10^{-4}$ (solid line). The peaks correspond to synchronization conditions for which $D^E \to \infty$ in the limit $D \to 0$

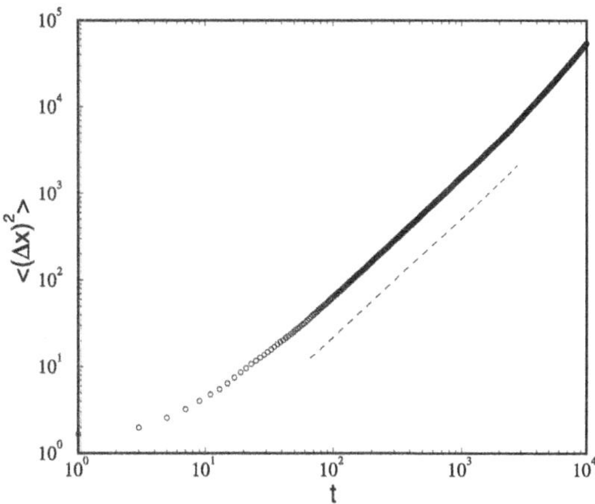

Fig. 8.6 Variance of the horizontal displacement $\langle x^2 \rangle$ for a set of trajectories that are solutions of (8.20) with $D = 0$ and $\epsilon = 1.1$. The dashed line represents the scaling law $\langle x^2 \rangle \sim t^{1.3}$ obtained from a fit of the numerical data

the *best fit* obtained from the numerical data and corresponds to the superdiffusive behavior $\langle x^2 \rangle \sim t^{1.3}$. Also in this case, the anomalous behavior is a consequence of the lack of decorrelation of the trajectories (that is, formally, of a diverging Lagrangian correlation time τ_L) which prevents the application of the general results from Sect. 7.4.

The probability density is not Gaussian, but unfortunately there is no analytical treatment that allows its expression to be determined.

Appendix: The Laplace Transform in Probability Theory

Given a function $f(x)$ with $x \geq 0$, the (direct) Laplace transform is defined as

$$\mathcal{L}f(s) = \int_0^\infty f(x)e^{-sx}dx. \tag{8.21}$$

If $f(x)$ is a probability density, then we can write the Laplace transform in the form

$$\mathcal{L}f(s) = E\left(e^{-sx}\right) \tag{8.22}$$

and it is therefore the moment generating function (if the moments exist). In fact, for s close to 0:

$$\mathcal{L}f(s) \simeq 1 - E(x)s + E(x^2)\frac{s^2}{2} + .. \tag{8.23}$$

$$E(x) = -\frac{d\mathcal{L}f(s)}{ds}\bigg|_{s=0}, \quad E(x^2) = \frac{d^2\mathcal{L}f(s)}{ds^2}\bigg|_{s=0}.$$

One expects that the Laplace transform is equivalent to the function $f(x)$; this is true under suitable (rather general) conditions; in fact, the inversion formula is:

$$f(x) = \frac{1}{2\pi i} \int_{x-i\infty}^{x+i\infty} e^{sx} \mathcal{L} f(s) ds.$$

The Laplace transform of the probability density enjoys properties very similar to those of the characteristic function; in particular, if the variables X and Y are independent, then, denoting by p_X, p_Y, and p_Z the probability densities of the variables X, Y, and $Z = X + Y$, we have

$$\mathcal{L}[p_X \star p_Y](s) = \mathcal{L}[p_X](s)\mathcal{L}[p_Y](s).$$

Let us now discuss the behavior at small values of s of the Laplace transform of a probability density with a power-law tail:

$$p(x) \sim \frac{A}{x^{1+\mu}}, \qquad A = \mu x_b^{\mu}. \tag{8.24}$$

If $\mu > 2$ then the first two moments exist and (8.23) holds. More interesting is the case with $0 < \mu < 2$.

Let us start by writing $\mathcal{L}[p](s)$ in the form:

$$\mathcal{L}[p](s) = 1 + \int_0^{\infty} p(x)\Big(e^{-sx} - 1\Big)dx$$

and divide the integral into two parts: from 0 to x^* and from x^* to ∞. Since $|e^{-sx} - 1| < sx$ for the first contribution we have

$$\Big|\int_0^{x^*} p(x)\Big(e^{-sx} - 1\Big)dx\Big| \leq sx^*.$$

For the second term, introducing the variable $y = sx$ and recalling the form of the probability density (8.24) we have

$$\int_{x^*}^{\infty} p(x)\Big(e^{-sx} - 1\Big)dx = \mu(x_b s)^{\mu} \int_{sx^*}^{\infty} (e^{-y} - 1)y^{-(1+\mu)}dy$$

integrating by parts and using the previous result we have

$$\mathcal{L}[p](s) = 1 - A_1 s^{\mu} + cs + .. \tag{8.25}$$

where $A_1 = x_b^{\mu}\Gamma(1 - \mu)$.

Exercises

8.1 Consider the standard map

$$I_{t+1} = I_t + K sin(\theta_t), \quad \theta_{t+1} = \theta_t + I_{t+1} \quad mod\, 2\pi.$$

Numerically calculate, as K varies, the diffusion coefficient

$$D = \lim_{t\to\infty} \frac{1}{2t} < (I_t - I_0)^2 >$$

where the average is over a large number of initial conditions and (I_0, θ_0) uniformly distributed in the square $[0, 2\pi] \times [0, 2\pi]$. Compare the results with the random phase approximation, in which it is assumed that the $\{\theta_t\}$ are independent and uniformly distributed in $[0, 2\pi]$: $D_{FA} = K^2/4$.

Note that for some values of K there is anomalous diffusion, that is, $< (I_t - I_0)^2 > \sim t^a$ with $a > 1$, which corresponds to an infinite value of the diffusion coefficient, see for example: P. Castiglione, A. Mazzino, P. Muratore-Ginanneschi and A. Vulpiani, "On strong anomalous diffusion", Physica D **134**, 75 (1999).

8.2 Consider the one-dimensional map

$$x_{t+1} = [x_t] + F(x_t - [x_t])$$

where [ ] denotes the integer part and

$$F(z) = \begin{cases} az & \text{if } 0 \le z < 1/2 \\ 1 + a(z-1) & \text{if } 1/2 \le z < 1 \end{cases}$$

Numerically calculate, as $a > 2$ varies, the diffusion coefficient

$$D = \lim_{t\to\infty} \frac{1}{2t} < (x_t - x_0)^2 >$$

where the average is taken over a large number of initial conditions and x_0 is uniformly distributed in $[0, 1]$.

You will find that $D(a)$ is a non-differentiable function, see for example R. Klages and J. R. Dorfman "Simple deterministic dynamical systems with fractal diffusion coefficients" Phys. Rev. E **59**, 5361 (1999).

Suggested Readings

For a rigorous treatment of infinitely divisible distributions and Lévy distributions, see the book by Gnedenko and that by Renyi already cited in Chaps. 2 and 3, and also:

B.V. Gnedenko, A.N. Kolmogorov, *Limit Distributions for Sums of Independent Random Variables* (Addison-Wesley, Reading, 1954)

For a not too technical discussion of Lévy distributions and their applications in physics:

F. Bardou, J.-P. Bouchaud, A. Aspect, C. Cohen-Tannoudji, *Lévy Statistics and Laser Cooling* (Cambridge University Press, Cambridge, 2001)

For a modern introduction to the theory of turbulence:

U. Frisch, *Turbulence* (Cambridge University Press, Cambridge, 1995)

A comprehensive introduction to the problem of anomalous transport is given in the review

J.P. Bouchaud, A. Georges, Anomalous Diffusion in disordered media: statistical mechanisms, models and physical applications. Phys. Rep. **195**, 127 (1990)

For the dispersion of pairs of particles in turbulence:

G. Boffetta, I.M. Sokolov, Relative dispersion in fully developed turbulence: The Richardson's Law and Intermittency Corrections. Phys. Rev. Lett. **88**, 094501 (2002);

For anomalous diffusion in deterministic systems:

P. Castiglione, A. Mazzino, P. Muratore-Ginanneschi, A. Vulpiani, On strong anomalous diffusion. Phys. D **134**, 75 (1999)

Chapter 9
Entropy, Information, and Chaos

This concluding chapter might seem like a digression largely unrelated to the theory of probability. However, as physicists, we considered it appropriate to include a brief general discussion on entropy, which, as we have already seen in Chap. 3, plays an important role in the theory of large deviations. This concept, originally introduced in thermodynamics by Clausius and developed in statistical mechanics—thanks especially to the contributions of Boltzmann—was later generalized by Wiener and Shannon in communication theory. Subsequently, the works of Kolmogorov and Sinai made it possible to use entropy for the characterization of the behavior of chaotic systems.

9.1 Entropy in Thermodynamics and Statistical Mechanics

In thermodynamics, entropy is a state function of a macroscopic system, defined such that its difference between two equilibrium states A and B is:

$$\Delta S_{AB} = S(B) - S(A) = \int_A^B \frac{dQ}{T},$$

where the integral is calculated along an arbitrary reversible transformation connecting A to B. In the case of n moles of a monoatomic ideal gas, denoting by V_A and T_A the volume and temperature of state A (and similarly V_B and T_B the volume and temperature of state B), a straightforward calculation shows:

$$\Delta S_{AB} = nR \ln \left[\frac{V_B}{V_A} \left(\frac{T_B}{T_A} \right)^{3/2} \right],$$

where R is the gas constant. Recalling that in a system with N particles, for the internal energy U we have $U = \frac{3}{2} N k_B T$ (where, denoting by N_A Avogadro's

G. Boffetta, A. Vulpiani, *Probability in Physics*, UNITEXT for Physics,
https://doi.org/10.1007/978-3-032-10407-6_9

number, $k_B = R/N_A$ is Boltzmann's constant), the previous formula can be rewritten as

$$S(N, V, U) = Nk_B \ln \left[\gamma_0 \left(\frac{U}{N} \right)^{3/2} \left(\frac{V}{N} \right) \right] \tag{9.1}$$

where γ_0 is a constant.

Equation (9.1) has a very instructive interpretation. Let us introduce the variables Δx and Δp defined as follows:

$$(\Delta x)^3 = \frac{V}{N}, \quad \frac{U}{N} = \frac{1}{2m} \left[\langle p_x^2 \rangle + \langle p_y^2 \rangle + \langle p_z^2 \rangle \right] = \frac{3}{2m} (\Delta p)^2;$$

Δx can be thought of as the typical interval available for each spatial coordinate of a particle; similarly, Δp is the typical variation of the momentum along a single direction.

We can rewrite (9.1) as

$$S(N, V, U) = k_B \ln \left[\frac{\Delta x \Delta p}{\gamma} \right]^{3N} \tag{9.2}$$

where $[\Delta x \Delta p / \gamma]^{3N}$ is the volume of phase space (of dimension $6N$) in units of γ^{3N}.

Equation (9.2) is nothing but a particular case of the "Boltzmann principle" that links thermodynamic properties to mechanical ones:

$$S = k_B \ln \Gamma_\Delta(E, V, N) \tag{9.3}$$

where

$$\Gamma_\Delta(E, V, N) = \int_{E<H<E+\Delta} \frac{d^{3N} x d^{3N} p}{N! h^{3N}}$$

is the volume of phase space contained between the hypersurfaces $H = E$ and $H = E + \Delta$. The factor $N!$ is introduced to avoid the Gibbs paradox, while the term h^{3N} in the classical context serves only to make Γ dimensionless.

Equation (9.3), which is engraved (with different notation) on the tomb of Ludwig Boltzmann in Vienna,[1] constitutes the fundamental bridge between thermodynamics and statistical mechanics in the microcanonical ensemble: by adding

[1] The engraved expression is:

$$S = k \log W$$

which, at least in this form, was not introduced by Boltzmann, but by Planck.

the "definition" of temperature

$$\frac{1}{T} = \frac{\partial S}{\partial E}$$

starting from (9.3), one can derive all of thermodynamics.

Another definition of entropy, due to Gibbs, is (in the canonical ensemble):

$$S = -k_B \int p(\mathbf{X}) \ln p(\mathbf{X}) d\mathbf{X} \tag{9.4}$$

where $\mathbf{X} \in R^{6N}$ and $p(\mathbf{X})$ is the canonical probability density. The reader can verify with an explicit calculation that in the limit $N \gg 1$ the two expressions are equivalent.[2]

9.2 Maximum Entropy Principle: Cornucopia or Pandora's Box?

In 1957, Jaynes proposed the maximum entropy principle as a rule for determining probabilities in situations where only partial information is available, for example, when some average values are known. Let us begin with the case of discrete variables; let p_i denote the (unknown) probability that the variable x takes the value $x^{(i)}$ (i = 1,2, …, N). If we know n average values:

$$\langle f_k \rangle = \sum_j p_j f_k(x^{(j)}) \quad k = 1, \ldots, n \tag{9.5}$$

(note that to ensure normalization, one of the average values will be $f_k = 1$, i.e., $\sum_j p_j = 1$) how do we determine the probabilities $\{p_j\}$? Jaynes' proposal is to maximize

$$S = -\sum_j p_j \ln p_j \tag{9.6}$$

subject to the constraints (9.5). It is easy to see that a multiplicative factor in the definition of S is irrelevant.

[2] S in the microcanonical ensemble is a function of E, V, and N, while in the canonical ensemble it depends on T, V, and N; Eqs. (9.3) and (9.4) are equivalent for $N \gg 1$ if $E = U$, where U is the average energy in the canonical ensemble.

Using the method of Lagrange multipliers, we obtain

$$\delta\Big[-\sum_j p_j \ln p_j - \sum_{k=1}^{n} \lambda_k \sum_j p_j f_k(x^{(j)})\Big] = 0$$

from which

$$p_j = \exp\left[-\sum_{k=1}^{n} \lambda_k f_k(x^{(j)})\right],$$

where, of course, the values of the Lagrange multipliers $\{\lambda_k\}$ are determined by the average values $\langle f_k \rangle$.

The entire procedure can easily be repeated for continuous variables; now S is defined as

$$S = -\int p(x) \ln p(x) dx, \tag{9.7}$$

yielding an expression entirely similar to (9.7):

$$p(x) = \exp\left[-\sum_{k=1}^{n} \lambda_k f_k(x)\right].$$

It is interesting that the maximum entropy principle, when applied to equilibrium statistical mechanics, very easily leads to the canonical probability distribution. In fact, denoting by $\mathbf{X}$ the $6N$-dimensional vector that determines the microscopic state, if we know the average value of the energy, the maximum entropy principle corresponds to

$$\delta\Big[-\int p(\mathbf{X}) \ln p(\mathbf{X}) d\mathbf{X} - \lambda_1 \int p(\mathbf{X}) E(\mathbf{X}) d\mathbf{X} - \lambda_2 \int p(\mathbf{X}) d\mathbf{X}\Big] = 0$$

from which the canonical probability density is

$$p(\mathbf{X}) = C\, e^{-\beta E(\mathbf{X})}$$

with C a normalization constant.

We have already seen in Chap. 2 that the entropy S defined as in (9.7) is not an intrinsic quantity: changing variables changes S. This has a rather serious (negative) consequence for the maximum entropy principle, since, once the average values are fixed, the probability density depends on the variables used.

One could therefore argue that the correct result obtained for the canonical ensemble is just a fortunate coincidence: using different variables, the result changes. We are back to the old Bertrand's paradox!

To overcome this problem, Jaynes proposed replacing the maximum entropy principle with a more sophisticated version: maximizing the relative entropy

$$\tilde{S} = -\int p(x) \ln \left[\frac{p(x)}{q(x)} \right] dx$$

where $q(x)$ is a given (known) probability density. Obviously, $\tilde{S}$ depends on $q(x)$; but, unlike S, it does not depend on the variables used. If a change of variable $x \to y = f(x)$ is performed, where the transformation is invertible, i.e., $f' \neq 0$, then $\tilde{S}$ does not change, see Exercise 9.1. At this point, the problem becomes selecting $q(x)$, and this is not much different from deciding on the "right variable" to use. For the canonical ensemble in statistical mechanics, the "right" q is the uniform distribution. In this case, it is not difficult to justify the choice; in fact, in the microcanonical ensemble, the probability density (as "suggested" by Liouville's theorem) is constant. But in the general case, it does not seem possible to appeal to any non-ad hoc criterion to indicate the "right" $q(x)$.

Jaynes is very explicit (we might call him an "anti-Boltzmann extremist"): statistical mechanics is not a part of physics, but rather a theory of statistical inference, a branch of epistemology. Supporters of the maximum entropy principle, following Jaynes, argue that, within the context of statistical mechanics, the method has at least a clear pedagogical advantage, since it allows one to obtain the same result in a very simple way which, with the "standard method," requires a certain amount of effort: discussion of the ergodic problem, justification of the microcanonical density, and then transition to the canonical ensemble. This point of view is not shared by authors who instead consider the connection between statistical mechanics and dynamics to be fundamental. It seems to us that the "advantage" is entirely a vicious circle. In this regard, we can recall a very caustic phrase by Lord Russell: "*To postulate what should be proved has many advantages: the same as theft over honest toil.*"

The debate on the principle of maximum entropy has been (and still is) very heated, at least in the case of continuous variables. The fundamental criticism, which we share, can be summarized in the old adage ex nihilo nihil; in other words, ignorance cannot be a source of inference.

9.3 Entropy and Information

The concept of entropy also plays a central role in the field of information theory. Let us consider the following problem: we have a probabilistic scheme A in which there are M events with probabilities $p_0, p_1, \ldots, p_{M-1}$, and we want to determine the degree of "uncertainty" of the scheme. In the case $M = 2$ the matter is rather simple: everyone would certainly agree that the scheme with $(p_0 = 0.2, p_1 = 0.8)$ is less uncertain than the one with $(p_0 = 0.3, p_1 = 0.7)$, and that the one with $(p_0 = 0.5, p_1 = 0.5)$ is more uncertain than the previous two. But already in the case

$M = 3$ things get complicated; for example, it is not easy to decide "barehanded" whether the scheme with $(p_0 = 0.2, p_1 = 0.3, p_2 = 0.5)$ is more or less uncertain than the one with $(p_0 = 0.15, p_1 = 0.4, p_2 = 0.45)$.

The problem was solved in 1948 by Shannon in his fundamental article on information theory. The function that unambiguously determines the uncertainty is

$$H = -\lambda \sum_{k=0}^{M-1} p_k \ln p_k,$$

where λ is an arbitrary positive constant. It can be shown (exercise for the reader) that H enjoys the following properties:

(a) H is continuous with respect to the $\{p_k\}$;
(b) H is maximal for $p_k = 1/M$ (hint: the function $f(x) = x \ln x$ is concave, i.e., $f'' > 0$);
(c) H is zero if there is a certain state, i.e., $p_k = \delta_{kj}$;
(d) H does not change if a state with zero probability is added $(p_{M+1} = 0)$;
(e) given two independent probabilistic schemes A and B, then

$$H(A, B) = H(A) + H(B);$$

(f) given two non-independent probabilistic schemes A and B where $p_k = P(\alpha_k)$ and $p_{ij} = P(\beta_i|\alpha_j)$, then

$$H(A, B) = H(A) + H(B|A)$$

where

$$H(B|A) = \sum_j p_j \Big[-\sum_i p_{ij} \ln p_{ij} \Big] \leq H(B). \tag{9.8}$$

The previous inequality has a clear intuitive meaning: the uncertainty of B knowing A is certainly no greater than that without knowing A.

Moreover, it can be shown (less simply) that the H defined in (9.8) is the only function (apart from the constant λ, which is fixed by the units of measurement) for which the previous 6 properties hold.

From now on, we will adopt the convention $\lambda = 1$ and the logarithms are natural (base e); computer scientists sometimes use base 2 logarithms.

9.3.1 Shannon Entropy

Let us consider an ergodic source that emits symbols $i_1, i_2, \ldots$ at discrete times, the symbols belong to an alphabet with M letters, each i_t takes values between 0 and $M-1$. We denote by W_N the words of length N; for example, if $M = 2$ (binary alphabet) the possible words W_1 are only two: $W_1 = 0$ and $W_1 = 1$, if $N = 2$ there are four possibilities $W_2 = (0, 0)$, $W_2 = (0, 1)$, $W_2 = (1, 0)$ and $W_2 = (1, 1)$, for $N = 3$ there are eight possible words $W_3 = (0, 0, 0)$, $W_3 = (0, 0, 1)$, $W_3 = (0, 1, 0)$, $W_3 = (1, 0, 0)$, $W_3 = (1, 1, 0)$, $W_3 = (1, 0, 1)$, $W_3 = (0, 1, 1)$ and $W_3 = (1, 1, 1)$.

Denoting by $P(W_N)$ the probability of the word W_N, we can introduce the block entropy of length N:

$$H_N = -\sum_{W_N} P(W_N) \ln P(W_N).$$

The quantity

$$h_N = H_N - H_{N-1}$$

has a well-defined meaning: it is the average uncertainty (in the sense of (9.8)) on the N-th symbol if the previous $N-1$ are known. In fact, noting that $W_N = (W_{N-1}, i_N)$, we can write $P(W_N) = P(W_{N-1})P(i_N|W_{N-1})$ and therefore

$$\begin{aligned} H_N = &-\sum_{W_{N-1}, i_N} P(W_{N-1})P(i_N|W_{N-1}) \ln P(W_{N-1}) \\ &-\sum_{W_{N-1}, i_N} P(W_{N-1})P(i_N|W_{N-1}) \ln P(i_N|W_{N-1}), \end{aligned}$$

summing over i_N in the first term and recalling that $\sum_{i_N} P(i_N|W_{N-1}) = 1$ we have

$$\begin{aligned} h_N = H_N - H_{N-1} &= -\sum_{W_{N-1}, i_N} P(W_{N-1})P(i_N|W_{N-1}) \ln P(i_N|W_{N-1}) = \\ &= \sum_{W_{N-1}} P(W_{N-1})[-\sum_{i_N} P(i_N|W_{N-1}) \ln P(i_N|W_{N-1})]. \end{aligned}$$

It is not difficult to show[3] that for $N \to \infty$, $h_N \to h_S$, called the Shannon entropy.

[3] Hint: Recall (9.8), and the well-known theorem on monotone sequences bounded above and below.

The simplest case is when the symbols $\{i_t\}$ are independent random variables:

$$h_S = -\sum_{k=0}^{M-1} p_k \ln p_k .$$

For Markov chains we have (exercise)

$$h_S = -\sum_{k,j} p_k P_{jk} \ln P_{jk} .$$

9.3.2 Shannon-McMillan Theorem

We now discuss an important theorem,[4] due to Shannon and McMillan, which shows the deep connection between Shannon entropy, statistical mechanics, and data compression.

Given an ergodic sequence $\{i_1, i_2, \ldots\}$ with Shannon entropy h_S, if $N \gg 1$ for the words W_N, two classes $\Omega_N^{(1)}$ and $\Omega_N^{(2)}$ can be identified with the following properties

$$\lim_{N\to\infty} \sum_{W_N \in \Omega_N^{(1)}} P(W_N) = 0,$$

while if $W_N \in \Omega_N^{(2)}$ then

$$P(W_N) \sim e^{-h_S N} \tag{9.9}$$

and furthermore

$$\lim_{N\to\infty} \sum_{W_N \in \Omega_N^{(2)}} P(W_N) = 1.$$

In other words, for large values of N there are two classes of words: the typical ones (in class $\Omega_N^{(2)}$) and the atypical ones (contained in $\Omega_N^{(1)}$). From (9.9) we have that the number of typical words is

$$\mathcal{N}(N) \sim e^{h_S N} . \tag{9.10}$$

[4] This theorem is closely related to the law of large numbers. The reader is invited to find a proof in the case of independent variables (recalling the law of large numbers) and for ergodic Markov chains.

If $h_S < \ln M$ then the number of typical words is much smaller than the number of possible words $M^N = e^{\ln MN}$.

Taking the logarithm of (9.10) and recalling that for $N \gg 1$, $H_N \simeq h_S N$, we obtain

$$H_N \sim \ln \mathcal{N}(N)$$

which is nothing but the "Boltzmann principle" in the context of information theory.

Note that the typical words are not at all the individually most probable ones (which are instead in class $\Omega_N^{(1)}$). To understand this, just consider the case of independent binary variables 0 and 1 with probabilities different from 1/2, for example 0.9 and 0.1; for each N the most probable word is clearly the one consisting of all zeros, that is, $(0, 0, \dots, 0)$ and has probability $(0.9)^N = e^{\ln(0.9)N}$. Instead, the typical words for $N \gg 1$, in accordance with the Shannon-McMillan theorem (which in this case is nothing but the law of large numbers), are those in which the number of zeros is 90%. For each of these words the probability is $(0.9)^{0.9N}(0.1)^{0.1N} = e^{-h_S N} \ll e^{\ln(0.9)N}$, where $h_S = -0.1\ln(0.1) - 0.9\ln(0.9) \simeq 0.324$ while $\ln(0.9) \simeq -0.105$.

Entropy and Data Compression

The Shannon-McMillan theorem has an important consequence in computer applications. Given a sufficiently long sequence $\{i_1, i_2, \dots, i_T\}$ in which the symbols are from an alphabet with M elements, one wants to "compress" the writing (i.e., the file) with a new series (for simplicity still with the same alphabet) $\{j_1, j_2, \dots, j_L\}$ in such a way that so that there is no loss of information (i.e., it is possible to reconstruct the original file exactly). Shannon showed that there is a limit on the value of L: if T is large enough, then the following inequality holds

$$\frac{L}{T} \geq \frac{h_S}{\ln M}.$$

In other words, the possibility of compressing data does not depend solely on our skill, but has an intrinsic limit determined by Shannon entropy: if the sequence has low entropy, it can be compressed a lot (if one is skilled), while if h_S is close to the maximum value $\ln M$ (in other words, the data are "very random"), then there is not much room for improvement, regardless of our ability.

9.3.3 Entropy and Chaos

Shannon entropy, with appropriate modifications, finds important use in the context of chaotic systems.

Let us consider a deterministic dynamical system; for simplicity, we limit ourselves to the discrete-time case:

$$\mathbf{x}_{t+1} = \mathbf{g}(\mathbf{x}_t), \tag{9.11}$$

once the initial condition $\mathbf{x}_0$ is assigned, we have a trajectory $\{\mathbf{x}_1, \mathbf{x}_1, \ldots, \mathbf{x}_T, \ldots\}$. Let us consider the case in which there is deterministic chaos (see Chap. 7), that is, the distance between two initially very close trajectories grows exponentially over time:

$$|\delta\mathbf{x}_t| \sim |\delta\mathbf{x}_0|\, e^{\lambda_1 t}$$

where $\lambda_1 > 0$ is the (maximum) Lyapunov exponent.

Let us now introduce a partition[5] $\mathcal{A}$ of the phase space (for example, into regular cells of size ϵ), to the trajectory $\{\mathbf{x}_1, \mathbf{x}_1, \ldots, \mathbf{x}_T, \}$ is associated a sequence of integers $\{i_1(\mathcal{A}), i_2(\mathcal{A}), \ldots, i_T(\mathcal{A})\}$ where $i_1(\mathcal{A})$ is the number corresponding to the cell containing $\mathbf{x}_1$, $i_2(\mathcal{A})$ is that determined by the cell containing $\mathbf{x}_2$, and so on. At this point, we can repeat the procedure discussed in information theory: calculate the $P(W_N)$, the H_N, and then in the limit $N \gg 1$ the Shannon entropy[6] $h_S(\mathcal{A})$ ($h_S(\epsilon)$ in the case of regular cells of size ϵ) which depends on the chosen partition. To overcome this and obtain a quantity that depends only on the system (9.11), one introduces the Kolmogorov-Sinai entropy:

$$h_{KS} = \sup_{\mathcal{A}} h_S(\mathcal{A}),$$

equivalently (and a bit more intuitively), in the case of a partition into cells of size ϵ, we have:

$$h_{KS} = \lim_{\epsilon \to 0} h_S(\epsilon).$$

The "physical" meaning of Kolmogorov–Sinai entropy is entirely analogous to Shannon entropy: h_{KS} measures the exponential proliferation of trajectories. Once the partition into cells is introduced, given the initial symbol $i_1(\epsilon)$, several sequences can be generated, since a given cell corresponds to different initial conditions. The

[5] The need to introduce a partition can arise for various reasons, for example, observation instruments with finite resolution, or for the requirements of an "effective description" of the problem; for example, for weather forecasts we might be satisfied (at least at a journalistic level) with a partition into 3 elements: good weather, bad weather, and uncertain weather.

[6] It is assumed that the system is ergodic and therefore the probabilities $P(W_N)$ can be calculated from a very long sequence, in terms of the frequencies of the words W_N.

number of sequences of length t (in practice, the trajectories with a coarse graining ϵ) grows as

$$\mathcal{N}(t) \sim e^{h_{KS} t}. \tag{9.12}$$

The nontrivial result is the validity of the previous equation independently of the value of ϵ if ϵ is small enough.

The entropy h_{KS} is determined by the exponential instability of the trajectories (sensitive dependence on initial conditions) of the chaotic system, in particular by the Lyapunov exponents. The connection between entropy and Lyapunov exponents can be understood simply in the case of low-dimensional systems (i.e., with only one positive Lyapunov exponent), for example, the Hénon map or the standard map. Imagine following the trajectories generated by the initial conditions contained in a circle of radius ϵ. As time passes, due to the stretching and folding mechanism present in chaotic systems, a sort of twisted filament will form, see Fig. 9.1.

Due to chaos, the length of this filament is $\sim \epsilon e^{\lambda_1 t}$, so the number of occupied cells grows as $\sim e^{\lambda_1 t}$. A moment of reflection convinces us that the number of

Fig. 9.1 Time evolution (from **a** to **d**) of a set of trajectories with initial condition in a circle of radius ϵ in a two-dimensional chaotic map on the torus

trajectories (with coarse graining ϵ) is also $\sim e^{\lambda_1 t}$ (it is enough to retrace the trajectory backwards starting from each final occupied cell). We therefore have

$$h_{KS} = \lambda_1 .$$

In systems with more than one positive Lyapunov exponent, the previous expression is generalized by Pesin's formula: the Kolmogorov-Sinai entropy is the sum of all positive Lyapunov exponents

$$h_{KS} = \sum_{i:\lambda_i > 0} \lambda_i .$$

9.4 Concluding Remarks

We have discussed various entropies: for statistical mechanics (Boltzmann and Gibbs), for information theory (Shannon), and for chaotic dynamical systems (Kolmogorov-Sinai). It is worth emphasizing that the common name entropy used in all cases is fully justified and not just at a formal level. In fact, although each has its own field of application, the various entropies share an important element: they count how "exponential proliferation" occurs: in statistical mechanics (9.3) the increase in the volume of phase space as the number of particles varies, in information theory the number of typical words as the length increases (9.10). Note that a consequence of the Shannon-McMillan theorem (9.10) is entirely analogous to the "Boltzmann principle." Similarly, for dynamical systems, the Kolmogorov-Sinai entropy is the exponential growth rate of the number of trajectories as time increases.

Exercises

9.1 Show that the relative entropy

$$\tilde{S}[p|q] = -\int p(x) \ln\left[\frac{p(x)}{q(x)}\right] dx$$

does not depend on the variables used, i.e., it is invariant under a change of variable $x \to y = f(x)$ where the transformation is invertible.

9.2 A particle moves with constant velocity on the segment $[0, L]$, when it hits the boundary at $x = L$ the velocity changes sign, whereas when it hits $x = 0$ a new velocity is drawn with probability density:

$$g(v) = \theta(v) v e^{-v^2/2},$$

where $\theta(v)$ is the step function, and each draw is independent of the others. Show that, by observing the particle for a very long time, for the probability densities one will observe a Gaussian for the velocity and a uniform distribution for the position:

$$p_V(v) = \frac{1}{\sqrt{2\pi}} e^{-v^2/2}, \quad p_X(x) = \frac{1}{L}.$$

This result is used in statistical mechanics simulations involving thermal baths, see for example R. Tehver, F. Toigo, J. Koplik and J.R. Banavar, "Thermal walls in computer simulations" Phys. Rev. E **57**, R17 (1998).

9.3 Show that in an ergodic Markov chain one has

$$H(x_n|x_0) = H(x_0|x_n),$$

where

$$H(x_0|x_n) = -\sum_{i,j} P(x_n = i)P(x_0 = j|x_n = i) \ln P(x_0 = j|x_n = i).$$

In other words, the present has a conditional entropy given knowledge of the past equal to that conditioned on knowledge of the future.

Discuss the limit $n \gg 1$.

9.4 Show that in an ergodic Markov chain in which detailed balance holds, one has

$$H(x_n|x_0 = i) = H(x_0|x_n = i),$$

for every i, where

$$H(x_n|x_0 = i) = -\sum_{j} P(x_0 = j|x_n = i) \ln P(x_0 = j|x_n = i).$$

Note that this property is stronger than that of the previous exercise.

9.5 Consider a Markov chain with two states:

$$P_{1\to 1} = p, \quad P_{1\to 2} = 1 - p, \quad P_{2\to 1} = 1, \quad P_{2\to 2} = 0.$$

Show that, for every $p \in (0, 1)$ the number of possible words $\mathcal{N}(N)$ of length N, for $N \gg 1$ grows as G^N where G is the golden ratio $G = (1 + \sqrt{5})/2$.

Hint: look for a relation between $\mathcal{N}(N)$, $\mathcal{N}(N-1)$ and $\mathcal{N}(N-2)$.

Note: the Markov chain in exercise *E7.4* corresponds to the case $p = 1/2$.

9.6 Consider the chaotic map

$$x_{t+1} = 3x_t \ \ mod\ 1,$$

numerically calculate the Shannon entropy relative to a partition with intervals $\epsilon = 1/m$ with $m = 2, 3, 4, \ldots$. Verify that for m a multiple of 3 you obtain the result predicted by Pesin's formula $h_{KS} = \ln 3$.

Suggested Readings

A beautiful book on statistical mechanics that discusses many aspects of entropy:

J.P. Sethna, *Statistical Mechanics: Entropy, Order Parameters and Complexity* (Oxford Univ. Press, Oxford, 2006); in particular Chap. 5

On L. Boltzmann as a man, physicist, and philosopher:

C. Cercignani, *Ludwig Boltzmann: The Man who Trusted Atoms* (Oxford University Press, Oxford, 1998)

On maximum entropy: the book that summarizes Jaynes' approach

E.T. Jaynes, *Probability Theory, The Logic of Science* (Cambridge University Press, Cambridge, 2003), the book was published posthumously, ed. by G. L. Bretthorst

For a critical discussion on maximum entropy:

J. Uffink, Can the maximum entropy principle be explained as a consistency requirement? Stud. History Philos. Mod. Phys. **26**, 223 (1995)

A small but great classic of information theory:

A.I. Khinchin, *On the Fundamental Theorems of Information Theory* (Dover Publications, New York, 1956)

On chaos and entropy:

M. Cencini, F. Cecconi, A. Vulpiani, *Chaos: From Simple Models to Complex Systems* (World Scientific, Singapore, 2009); in particular Chap. 8

Chapter 10
Appendix: Some Useful Results and Complements

This appendix has been introduced for completeness, in order to summarize some results already known to the reader, some useful results, and additional complements.

10.1 Marginal and Conditional Probability Densities

In Chap. 1, when discussing the case of a scalar random variable, we introduced the distribution function $F_X(x) = P(X < x)$. For random variables in higher dimensions, there is an obvious generalization. For simplicity of notation, let us begin with the case of 2 variables by defining the joint distribution function

$$F_{X,Y}(x, y) = P(X < x, Y < y). \tag{10.1}$$

Naturally, for each fixed value of y, $F_{X,Y}(x, y)$ is a non-decreasing function of x; similarly, for each x, $F_{X,Y}(x, y)$ is a non-decreasing function of y.

In the case where $F_{X,Y}(x, y)$ is differentiable, one can introduce the joint probability density

$$p_{X,Y}(x, y) = \frac{\partial^2}{\partial x \partial y} F_{X,Y}(x, y). \tag{10.2}$$

The meaning of $p_{X,Y}(x, y)$ is clear:

$$P((x, y) \in A) = \int_A p_{X,Y}(x', y') dx' dy'.$$

G. Boffetta, A. Vulpiani, *Probability in Physics*, UNITEXT for Physics,
https://doi.org/10.1007/978-3-032-10407-6_10

If one agrees to consider even singular functions (for example, the Dirac delta) as probability densities, then it is not necessary to treat discrete and continuous random variables separately.

One can introduce the marginal distribution functions

$$F_X(x) = P(X < x) = F_{X,Y}(x, \infty) \ , \quad F_Y(y) = P(Y < y) = F_{X,Y}(\infty, y) \ ,$$

and the marginal probability densities:

$$p_X(x) = \int p_{X,Y}(x, y)dy, \quad p_Y(y) = \int p_{X,Y}(x, y)dx, \tag{10.3}$$

obviously

$$p_X(x) = \frac{d}{dx}F_X(x), \quad p_Y(y) = \frac{d}{dy}F_Y(y).$$

10.1.1 Conditional Probability Density

Given two intervals $a = [x_1, x_2]$ and $b = [y_1, y_2]$, the probability of having the variable Y in b conditioned on X being in a is a well-defined quantity:

$$P(Y \in b|X \in a) = \frac{P(X \in a, Y \in b)}{P(X \in a)}.$$

It is interesting to note that the conditional probability introduced above is well defined even in the limit $P(X \in a) \to 0$. Let us consider the case $x_2 = x_1 + \Delta x$, $y_2 = y_1 + \Delta y$ with Δx and Δy small, we have

$$P(Y \in b|X \in a) = \frac{p_{X,Y}(x_1, y_1)\Delta x \Delta y}{p_X(x_1)\Delta x},$$

we can therefore coherently define a conditional probability density:

$$p_{Y|X}(y|x) = \frac{p_{X,Y}(x, y)}{p_X(x)} \tag{10.4}$$

such that

$$P(Y \in b|X = x_1) = p_{Y|X}(y_1|x_1)\Delta y.$$

Note that (10.4) has the "same form" as in the discrete case.[1]

In the case of independent variables, we have

$$p_{X,Y}(x, y) = p_X(x)p_Y(y), \quad p_{Y|X}(y|x) = p_Y(y), \quad p_{X|Y}(x|y) = p_X(x).$$

As an example, let us consider a particular case of the bivariate Gaussian distribution (which will be discussed again later):

$$p_{X,Y}(x, y) = \frac{1}{2\pi\sqrt{1-\rho^2}} exp - \frac{x^2 + y^2 - 2\rho xy}{2(1-\rho^2)},$$

where ρ is called the correlation coefficient between X and Y, $|\rho| \le 1$, and obviously if $\rho = 0$, X and Y are independent. As an exercise, the reader can verify that, for any value of ρ, the marginal distributions are Gaussian, as are the conditional distributions.

10.1.2 Three or More Variables

In the case of 3 variables, we have the joint distribution function

$$F_{X,Y,Z}(x, y, z) = P(X < x, Y < y, Z < z)$$

and the probability density

$$p_{X,Y,Z}(x, y, z) = \frac{\partial^3}{\partial x \partial y \partial z} F_{X,Y,Z}(x, y, z).$$

[1] Denoting by $P_{i,j}$ the probability $P(x = x_i, y = y_j)$, where $x_1, x_2 \ldots$ are the possible discrete values of the variable X (and similarly for Y), the marginal probabilities are

$$P_i^X = \sum_j P_{i,j}, \quad P_j^Y = \sum_i P_{i,j}$$

and the conditional probabilities (with obvious meaning of the symbols)

$$P_{i|j}^{X|Y} = \frac{P_{i,j}}{P_j^Y}, \quad P_{j|i}^{Y|X} = \frac{P_{i,j}}{P_i^X}.$$

We can introduce 3 probability densities that depend on a single variable:

$$p_X(x) = \int\int p_{X,Y,Z}(x, y, z)dydz$$

and similarly $p_Y(y)$, $p_Z(z)$; and also 3 densities that depend on 2 variables:

$$p_{X,Y}(x, y) = \int p_{X,Y,Z}(x, y, z)dz$$

and similarly $p_{Y,Z}(y, z)$, $p_{X,Z}(x, z)$.

We have two different types of conditional densities; if we condition with respect to one variable (for example z) we have

$$p_{\{X,Y\}|Z}(x, y|z) = \frac{p_{X,Y,Z}(x, y, z)}{p_Z(z)},$$

if we condition with respect to two variables (for example y and z) we have

$$p_{X|\{Y,Z\}}(x|\{y, z\}) = \frac{p_{X,Y,Z}(x, y, z)}{p_{Y,Z}(y, z)}.$$

The case with more variables can obviously be generalized without any difficulty.

10.2 Mean Values

Given a random variable X with probability density $p_X(x)$, the mean value of a function $f(x)$ is defined as

$$< f(x) >= E(f(x)) = \int f(x)p_X(x)dx; \tag{10.5}$$

in the multidimensional case

$$< f(\mathbf{x}) >= E(f(\mathbf{x})) = \int f(\mathbf{x})p_{X_1,\ldots,X_N}(x_1, \ldots, x_N)dx_1 \ldots dx_N.$$

Particularly important are the cases with $f(x) = x$ and $f(x) = x^2$, the variance σ_x^2 is defined as

$$\sigma_x^2 =< x^2 > - < x >^2 =< (x- < x >)^2 > .$$

Let us denote by y the sum $a_1x_1 + \ldots + a_Nx_N$ where $x_1, \ldots, x_N$ are random variables and $a_1, \ldots, a_N$ are real constants, the following properties are easily proven (left to the reader):

$$E(y) = E(a_1x_1 + \ldots + a_Nx_N) = a_1E(x_1) + \ldots + a_NE(x_N),$$

$$\sigma_y^2 = \sum_{j=1}^{N} a_j^2\sigma_{x_j}^2 + 2\sum_{j<n} a_ja_nE\Big((x_j - E(x_j))(x_n - E(x_n))\Big).$$

In the case of independent variables:

$$E\Big(\prod_{j=1}^{N} f_j(x_j)\Big) = \prod_{j=1}^{N} E(f_j(x_j)), \;\; \sigma_y^2 = \sum_{j=1}^{N} a_j^2\sigma_{x_j}^2.$$

10.2.1 A Frequently Useful Inequality

Let us consider the case in which $f(x)$ is a concave function, that is, $f''(x) \geq 0$, the following inequality (Jensen's inequality) holds:

$$E(f(x)) \geq f(E(x)). \tag{10.6}$$

The proof is very simple. From the property of concavity we have that for every x_0 the following relation holds

$$f(x) \geq f(x_0) + f'(x_0)(x - x_0).$$

Taking the mean of the previous inequality and setting $x_0 = E(x)$ we obtain (10.6). Similarly, if $f(x)$ is a convex function, that is, $f''(x) \leq 0$, we have

$$E(f(x)) \leq f(E(x)).$$

From Jensen's inequalities, useful relations of the type follow:

$$E(x^2) \geq [E(|x|)]^2, \;\; E(x^4) \geq [E(x^2)]^2$$

$$E\Big[e^{ax}\Big] \geq e^{aE(x)}, \;\; E(\ln x) \leq \ln E(x).$$

10.2.2 Conditional Mean Values

For simplicity of notation, let us consider the case of two variables X and Y with joint probability density $p_{X,Y}(x, y)$, we define the expectation value of a function

$f(x)$ conditioned on a given value y as

$$E(f(x)|y) = \int f(x) p_{X|Y}(x|y) dx,$$

it is possible to express the unconditional mean value $E(f(x))$ as follows

$$E(f(x)) = E_Y(E(f(x)|y)) = \int E(f(x)|y) p_Y(y) dy, \tag{10.7}$$

where E_Y indicates that the mean value is calculated with the marginal distribution of the variable Y. The proof is immediate: writing $p_{X,Y}(x, y) = p_{X|Y}(x|y)p_Y(y)$ we have

$$E(f(x)) = \int\int f(x) p_{X,Y}(x, y) dx dy = \int \Big[\int f(x) p_{X|Y}(x|y) dx\Big] p_Y(y) dy =$$

$$\int E(f(x)|y) p_Y(y) dy = E_Y(E(f(x)|y)).$$

10.2.3 From Moments to the Probability Density

Of course, given the probability density $p_X(x)$, one can calculate the moments $M_n =< x^n >$; one may ask whether it is possible to determine $p_X(x)$ starting from the moments. Since we can write the characteristic function in the form

$$\phi_X(t) = 1 + \sum_{n=1} \frac{M_n}{n!} (it)^n,$$

recalling the theorems from complex analysis on power series, it is clear that if the moments M_n do not grow too quickly with n, for example $M_n < C^n$, then from the knowledge of all the moments it is possible to uniquely determine $\phi_X(t)$ and therefore $p_X(x)$.

However, if M_n grows too quickly with n, it is not possible to reconstruct $p_X(x)$ from $\{M_n\}$; in other words, it is possible to have two different probability densities with the same moments.

As an example, we can consider the lognormal distribution defined for $x > 0$:

$$p_{LN}(x) = \frac{1}{x\sqrt{2\pi}} e^{-(\ln x)^2/2},$$

and the function

$$f(x) = \frac{1}{x\sqrt{2\pi}}\Big[1 + b\,sin(\ln(x))\Big]e^{-(\ln x)^2/2}.$$

It is easy to verify (exercise for the reader) that if $|b| < 1$ then $f(x) \geq 0$ and moreover

$$\int_0^\infty f(x)dx = 1,$$

so $f(x)$ is a probability density and is clearly different from $p_{LN}(x)$. The calculation of the moments (left as an exercise) shows that

$$\int_0^\infty f(x)x^n dx = \int_0^\infty p_{LN}(x)x^n dx = e^{n^2/2},$$

so we have two different probability densities with the same moments.

10.2.4 Cumulants

Recall from Chap. 3 that the cumulants are defined from the cumulant generating function

$$L(q) = \ln\left[E\left(e^{qx}\right)\right].$$

The characteristic function $\phi_X(t)$ can be written in terms of the moments (if they exist). It can be useful to express $\phi_X(t)$ in a different way with the cumulants $c_1, c_2, \ldots$:

$$\ln \phi_X(t) = \sum_{n=1} \frac{c_n}{n!}(it)^n.$$

Naturally, the cumulants are closely related to the moments; a straightforward calculation shows:

$$c_1 = M_1, \;\; c_2 = M_2 - M_1^2,$$

$$c_3 =< (x - M_1)^3 >, \;\; c_4 =< (x - M_1)^4 > -3(c_2)^2 \;\; \ldots.$$

It is immediate to verify that if $p_X(x)$ is Gaussian, for $n > 3$ we have $c_n = 0$.

10.3 Some Notable Distributions

10.3.1 Binomial Distribution

This distribution, although elementary, plays an important role in probability theory. Let us consider the variable $y_N = x_1 + x_2 + \ldots + x_N$, where the $\{x_i\}$ are i.i.d. and take the value 1 or 0 with probability p and $1 - p$ respectively. The probability of having $y_N = k$ is:

$$P_N(k) = C_{N,k}\, p^k (1-p)^{N-k}$$

where $C_{N,k}$ is the number of ways (combinations) in which k indistinguishable objects can be arranged in a sequence of length N. The well-known expression[2] for $C_{N,k}$ is

$$C_{N,k} = \frac{N!}{k!(N-k)!}, \tag{10.8}$$

and therefore

$$P_N(k) = \frac{N!}{k!(N-k)!}\, p^k (1-p)^{N-k}. \tag{10.9}$$

The calculation of the mean and variance of y_N is elementary: since the $\{x_i\}$ are i.i.d. we have

$$E(y_N) = NE(x) = Np, \;\; \sigma^2_{y_N} = N\sigma^2_x = Np(1-p),$$

obviously the same results can be obtained starting from (10.9).

From the binomial distribution, in the limit $N \gg 1$ and finite p, one obtains the Gaussian distribution; for details of the derivation based on Stirling's approximation, see Sect. 3.4 of Chap. 3.

Another interesting case is that with $N \gg 1$ and $p = \lambda/N$.

---

[2] Equation (10.8) is an elementary formula from combinatorics. For completeness, let us derive it starting from the recurrence relation

$$C_{N+1,k} = C_{N,k} + C_{N,k-1}$$

which expresses in formulas the observation that binary sequences of length $N+1$ containing k ones are those of length N with k ones, followed by a 0, plus those of length N with $k-1$ ones, followed by a 1. Noting that $C_{N,1} = N$ and $C_{N,0} = 1$, by induction we obtain the result (10.8).

10.3.2 Poisson Distribution

Let us consider the limiting case of the binomial distribution in which $N \gg 1$, $p = \lambda/N$ with $\lambda = O(1)$ and $k \ll N$. Noting that

$$\frac{N!}{k!(N-k)!} = \frac{N(N-1)\dots(N-k+1)}{k!} \simeq \frac{N^k}{k!}$$

and

$$(1-p)^{N-k} = \left(1 - \frac{\lambda}{N}\right)^{N-k} \simeq \left(1 - \frac{\lambda}{N}\right)^{N} \simeq e^{-\lambda},$$

from (10.9) we have

$$P(k) = \frac{\lambda^k}{k!} e^{-\lambda} \quad k = 0, 1, 2, \dots \tag{10.10}$$

i.e., an expression independent of N. Note that the probability is correctly normalized: $P(0) + P(1) + \dots = 1$. The meaning of λ is clear from the calculation of $E(k) =< k >$:

$$E(k) = \sum_{k=0}^{\infty} kP(k) = \sum_{k=0}^{\infty} k \frac{\lambda^k}{k!} e^{-\lambda} = e^{-\lambda} \lambda \frac{\partial}{\partial \lambda} \sum_{k=0}^{\infty} \frac{\lambda^k}{k!} = \lambda,$$

with a similar calculation we have

$$\sigma^2 = E(k^2) - E(k)^2 = \lambda.$$

A simple but interesting application for statistical mechanics of the Poisson distribution is the probability of finding k particles in a small region of volume ΔV in a container of volume $V \gg \Delta V$, containing a very large number N of particles. Assuming that the particles are distributed uniformly, then the probability that a given particle is contained in a region of volume ΔV is $p = \Delta V/V$. Introducing the density $\rho = N/V$, we can write $p = \rho \Delta V/V =< k > /N$, where the average number of particles is $< k >= Np$. We therefore have that in the grand canonical ensemble the probability of having k particles is given by (10.10) with $\lambda =< k >$.

10.3.3 Pearson's χ^2 Distribution

Let $x_1, \dots, x_N$ be i.i.d. variables with Gaussian probability density with zero mean and unit variance, we have

$$p_{X_1,\dots,X_N}(x_1, \dots, x_N) = \sqrt{\frac{1}{(2\pi)^N}}\, exp - \frac{1}{2}\sum_{j=1}^{N} x_j^2.$$

Let us consider the variable

$$\chi_N^2 = \sum_{j=1}^{N} x_j^2$$

using (2.16) for the variable $y = \chi_N^2$, and recalling the definition of Euler's gamma function (see below), we have

$$p_Y(y) = \frac{y^{(N/2-1)}}{2^{N/2}\Gamma(N/2)} e^{-\frac{y}{2}}. \tag{10.11}$$

The above function is called the Pearson χ^2 distribution for N degrees of freedom.

Similarly, for the variable $z = \chi_N$ we have

$$p_Z(z) = 2\frac{z^{N-1}}{2^{N/2}\Gamma(N/2)} e^{-\frac{z^2}{2}}. \tag{10.12}$$

The probability distribution of χ^2 (or equivalently that for χ) plays an important role in the treatment of experimental data. In fact, it is natural to expect that the difference between an experimental observation and the "true" value is a Gaussian variable.

10.3.4 More on the Poisson Distribution

Let us consider $x_1, \dots, x_N$ positive i.i.d. variables with exponential probability density

$$p_1(x) = \lambda\, e^{-\lambda x}. \tag{10.13}$$

Let us calculate the probability density of $z = x_1 + x_2$:

$$p_2(z) = \int p_1(x)p_1(z-x)dx = (p_1 * p_1)(z) = \lambda^2\, z\, e^{-\lambda z}.$$

It is easy to repeat the calculation for $z = x_1 + \ldots + x_N$:

$$p_N(z) = \frac{\lambda^N}{(N-1)!} z^{N-1} e^{-\lambda z}, \quad (10.14)$$

which is nothing but (10.11) with $N/2$ replaced by N and $1/2$ replaced by λ.

The probability distribution (10.14) plays an important role in the statistics of industrial accidents. In fact, as an empirical fact, (10.13) satisfactorily describes the probability density of the failure time of a machine (and also the lifetime of a light bulb); therefore, if we imagine immediately replacing the machine after the accident, (10.14) represents the probability distribution of the lifetime of N machines.

Let us now ask what is the probability $P_T(n)$ that in the interval $(0, T)$ there are n failures. Let x_j be the lifetime of the j-th machine, there are n failures in $(0, T)$ if

$$\sum_{j=1}^{n} x_j < T \ and \ \sum_{j=1}^{n+1} x_j > T,$$

therefore

$$P_T(n) = \int_0^T p_n(x)dx - \int_0^T p_{n+1}(x)dx$$

where $p_n(x)$ and $p_{n+1}(x)$ are given by (10.14). Integrating by parts the first integral, we obtain

$$P_T(n) = \frac{(\lambda T)^n}{n!} e^{-\lambda T},$$

so the number of failures in an interval of duration T is a Poisson variable with $<n>= \lambda T$.

10.3.5 Multidimensional Distribution of Gaussian Variables

We have encountered several times the Gaussian probability distribution with mean m and variance σ^2:

$$p_X(x) = \frac{1}{\sqrt{2\pi\sigma^2}} exp\left[-\frac{1}{2\sigma^2}(x-m)^2\right],$$

its generalization to the case of N independent variables $x_1, \ldots, x_N$, each with mean m_j and variance σ_j^2, is immediate:

$$p_{X_1,\ldots,X_N}(x_1, \ldots, x_N) = \prod_n \frac{1}{\sqrt{2\pi\sigma_n^2}} exp\left[-\sum_n \frac{1}{2\sigma_n^2}(x_n - m_n)^2\right].$$

Let us consider new variables $y_1, \ldots, y_N$ expressible as linear combinations of $x_1, \ldots, x_N$:

$$\mathbf{y} = \mathcal{B}\mathbf{x} + \mathbf{a}$$

where $\mathcal{B}$ is a matrix with nonzero determinant. It is easy to see that the probability distribution of $\mathbf{y}$ is of the form:

$$p_{\mathbf{Y}}(y_1, \ldots, y_N) = \sqrt{\frac{|det\mathcal{A}|}{(2\pi)^N}} exp\left[-\frac{1}{2}\sum_{i,j}(y_i - b_i)(y_j - b_j)A_{ij}\right], \quad (10.15)$$

where $\mathcal{A}$ is a symmetric positive definite matrix (i.e., with positive eigenvalues) and $\{b_j\}$ are the mean values of $\{y_j\}$. It can be easily shown[3] that:

$$< (y_i - b_i)(y_j - b_j) >= \left[\mathcal{A}^{-1}\right]_{ij}.$$

Equation (10.15) is called the multivariate Gaussian.

In the case $N = 2$ where $< x_1 >=< x_2 >= 0$ and $\sigma_1 = \sigma_2 = 1$ the most general form of the bivariate Gaussian is:

$$p_{X_1,X_2}(x_1, x_2) = \frac{1}{2\pi\sqrt{1-\rho^2}} exp\left[-\frac{x_1^2 + x_2^2 - 2\rho x_1 x_2}{2(1-\rho^2)}\right], \quad (10.16)$$

where ρ is the correlation coefficient between x_1 and x_2: $< x_1 x_2 >= \rho$ and $|\rho| \leq 1$. The general case of two variables y_1 and y_2, with mean values m_1 and m_2, and variances σ_1^2 and σ_2^2, with a bivariate Gaussian distribution is easily obtained from (10.16) with the change of variables:

$$y_1 = m_1 + \sigma_1 x_1 \quad y_2 = m_2 + \sigma_2 x_2$$

[3] Just change variables:

$$\mathbf{y} \to \mathbf{z} = \mathcal{C}(\mathbf{y} - \mathbf{b})$$

so that $z_1, \ldots, z_N$ are independent, calculate $< z_j^2 >$ and then return to $< (y_i - b_i)(y_j - b_j) >$.

thus obtaining:

$$p_{Y_1,Y_2}(y_1, y_2) = \frac{1}{2\pi\sigma_1\sigma_2\sqrt{1-\rho^2}} \times$$

$$\times\, exp\left\{-\frac{1}{2(1-\rho^2)}\left[\left(\frac{y_1-m_1}{\sigma_1}\right)^2 - 2\rho\left(\frac{y_1-m_1}{\sigma_1}\right)\left(\frac{y_2-m_2}{\sigma_2}\right) + \left(\frac{y_1-m_2}{\sigma_2}\right)^2\right]\right\},$$

and

$$\rho = \frac{1}{\sigma_1\sigma_2} < (y_1 - m_1)(y_2 - m_2) > .$$

10.4 Euler Gamma Function and Stirling's Approximation

In the theory of probability, and in statistical mechanics, the Euler gamma function often appears

$$\Gamma(x) = \int_0^\infty t^{x-1} e^{-t} dt.$$

We limit ourselves to the case with x real and positive; by integrating by parts it is immediate to verify that

$$\Gamma(1) = 1, \quad \Gamma(x+1) = x\Gamma(x),$$

so for integer values of n we have $\Gamma(n+1) = n!$.

The gamma function appears in the calculation of the volume of hyperspheres of dimension D:

$$V_D(R) = \int_{\sum_{j=1}^D x_j^2 \leq R^2} dx_1 \ldots .dx_D = C_D R^D,$$

where C_D is the volume of the D-dimensional hypersphere of unit radius. The determination of C_D in terms of the gamma function can be obtained as follows. Let us consider the integral

$$I_D = \int e^{-\sum_{j=1}^D x_j^2} dx_1 \ldots .dx_D = \pi^{\frac{D}{2}},$$

noting that $dV_D(R) = DC_D R^{D-1} dR$, we have

$$I_D = DC_D \int_0^\infty R^{D-1} e^{-R^2} dR,$$

with the change of variables $x = R^2$ we get

$$I_D = \frac{D}{2} C_D \int_0^\infty x^{D/2-1} \, e^{-x} dx = \frac{D}{2} C_D \Gamma(\frac{D}{2}),$$

from which

$$C_D = \frac{\pi^{\frac{D}{2}}}{\frac{D}{2}\Gamma(\frac{D}{2})} = \frac{\pi^{\frac{D}{2}}}{\Gamma(\frac{D}{2}+1)}.$$

It is easy to see that $N! = \Gamma(N+1)$ grows very rapidly with N (for example $5! = 120$, $20! \simeq 2.432 \times 10^{18}$) and it is therefore important to have an (even approximate) expression for $N!$ for large N. The answer to this problem is Stirling's approximation.

We write $N!$ in the form

$$N! = \Gamma(N+1) = \int_0^\infty t^N \, e^{-t} dt,$$

introducing the variable $z = t/N$ we have

$$N! = N^{N+1} \int_0^\infty e^{N(\ln z - z)} dz. \tag{10.17}$$

For large N the integral can be (approximately) calculated using the Laplace method.

10.4.1 The Laplace Method

Let us consider the integral

$$I = \int_a^b e^{Nf(x)} dx \tag{10.18}$$

where $N \gg 1$ and $f(x)$ has a quadratic maximum at $x_0 \in [a, b]$. It is easy to see, at least at a non-rigorous level, that the dominant contribution to I comes from the region around x_0, so by approximating $f(x)$ with its Taylor expansion:

$$f(x) \simeq f(x_0) - \frac{1}{2} |f_0''| (x - x_0)^2,$$

where f_0'' denotes the second derivative at x_0, we have

$$I \simeq e^{Nf(x_0)} \int_a^b e^{-\frac{N}{2}|f_0''|(x-x_0)^2} dx,$$

since $x_0 \in [a, b]$, and $N \gg 1$, we can approximate I as

$$I \simeq e^{Nf(x_0)} \int_{-\infty}^{\infty} e^{-\frac{N}{2}|f_0''|(x-x_0)^2} dx$$

since the contributions for $x < a$ and $x > b$ are exponentially small. At this point, recalling the well-known formula for the Gaussian integral

$$\int_{-\infty}^{\infty} e^{-ax^2} dx = \sqrt{\frac{\pi}{a}},$$

we obtain

$$I \simeq e^{Nf(x_0)} \sqrt{\frac{2\pi}{N|f''(x_0)|}}.$$

Using the result just obtained for (10.17) with $f(x) = \ln x - x$, we get:

$$N! = \Gamma(N+1) \simeq N^N e^{-N} \sqrt{2\pi N},$$

this approximation (called Stirling's) is very accurate even for small values of N, for example for $N = 2, 3, 20, 40$ and 100 with Stirling's formula for $N!$ we obtain $1.91, 5.95, 2.42 \times 10^{18}, 8.14 \times 10^{47}$ and 9.32×10^{157}, to be compared with the exact values $2, 6, 2.43 \times 10^{18}, 8.16 \times 10^{47}$ and 9.33×10^{157} respectively.

If a more accurate expression is desired, one can use

$$N! = \Gamma(N+1) \simeq N^N e^{-N} \sqrt{2\pi N}\Big[1 + \frac{1}{12N} + \frac{1}{144N^2} + O(N^{-3})\Big].$$

10.5 The Contribution of a Great Mathematician to Probability in Genetics: An Elementary Calculation

G.H. Hardy, a leading figure in twentieth-century mathematics, famous for his contributions to number theory, saw mathematics as a science completely detached from the real world and boasted that none of his works had ever had any practical relevance. Nevertheless, in a great irony, his name is today much more widespread in biology textbooks for his (occasional) two-page contribution with an elementary probability calculation applied to genetics, than in mathematics books for his sophisticated works in analysis.

The problem solved by Hardy is the following:[4]

Why does a dominant genetic trait not increase, over the course of generations, its frequency so that the recessive trait asymptotically disappears?

Let us consider the three possible pairs of Mendelian traits (genotypes) in the case where there are only two types of genes A and S: AA, AS, and SS; let AA be the purely dominant trait, AS the heterozygote, and SS the purely recessive trait. Assume that in the initial generation the probabilities of AA, AS, and SS are p_0, $2q_0$, and r_0, with $p_0+2q_0+r_0 = 1$. Suppose that pairings occur without preference and that the probability of the trait does not depend on sex; under these assumptions, the calculation of the probabilities of AA, AS, and SS after one generation is a simple exercise.[5] You can have AA only with the combination of AA with AA; AA with AS and AS with AS, so

$$p_1 = p_0^2 + 2q_0 p_0 + q_0^2 = (q_0 + p_0)^2 \tag{10.19a}$$

similarly, we obtain

$$2q_1 = 2(r_0 + q_0)(q_0 + p_0), \quad r_1 = (q_0 + r_0)^2. \tag{10.19b}$$

Let us introduce the concept of gene frequency for gene A, as the ratio between the number of A genes present in a (large) population containing N individuals, and the total number of genes. At the initial generation we have:

$$f_0(A) = \frac{2p_0N + 2q_0N}{2N} = p_0 + q_0, \quad f_0(S) = r_0 + q_0,$$

obviously $f_0(A) + f_0(S) = 1$. Equations (10.19) can be rewritten as:

$$p_1 = f_0(A)^2 \ , \ 2q_1 = 2f_0(A)f_0(S) \ , \ r_1 = f_0(S)^2.$$

After one generation we have

$$f_1(A) = p_1 + q_1 = (q_0+p_0)^2 + (q_0+p_0)(q_0+r_0) = f_0(A)[f_0(A)+f_0(S)] = f_0(A) \ ,$$

[4] The result was obtained independently also by the physician W. Weinberg and in biology it is known as the *Hardy-Weinberg equilibrium*. It seems that the origin of the work was a dinner discussion at Trinity College, Cambridge, with the geneticist (and his cricket companion) R.O. Punnet. Hardy solved the problem immediately, but considered the result so trivial as not to deserve publication; here is how the article begins: "I am reluctant to intrude in a subject of which I have no thorough knowledge, and I should have expected that the very simple observation which I wish to make was familiar to biologists."

[5] Hardy did not consider it important to go into details decidedly too elementary for the great number theory expert *A little mathematics of the multiplication table type is enough to show.*

therefore, for $n \geq 1$

$$p_n = p_1 \ , \ 2q_n = 2q_1 \ , \ r_n = r_1 \ .$$

As generations pass, from the second onward, the relative frequencies of the genotypes AA, AS, and SS, contrary to what intuition might erroneously suggest, do not change.

Obviously, in reality, the assumptions underlying this calculation (for example, random mating) are not always entirely realistic. Deviations from the predictions are an indication that external factors have intervened, altering the Hardy-Weinberg equilibrium.

10.6 Statistics of Extreme Events

A problem of obvious practical interest is the statistical behavior of the extreme values of random variables. For example, consider river floods: if one wants to plan the construction of levees (and thus decide the height of protective barriers), one must have a good understanding of the probability distribution of the maximum flow rate in a given time interval.

Let us consider a random variable X with distribution function $F_X(x)$, that is:

$$F_X(x) = P(X < x).$$

If X is the maximum flow rate of a river in a year, then $F_X(x)$ is the probability that the maximum flow rate does not exceed the value x. A useful quantity is

$$T(x) = \frac{1}{1 - F_X(x)},$$

which has a rather clear interpretation: $T(x)$ is the mean time (in the example of river floods, the time in years) until $X > x$. If n observations are made, the relation

$$n(1 - F_X(u_n)) = 1$$

defines the variable u_n, which can be interpreted as the "typical maximum value" in n observations.

Let us consider n i.i.d. random variables $x_1, \ldots, x_n$ with distribution function $F_X(x)$, and probability density $p_X(x) = dF_X(x)/dx$. We denote by y_n the maximum among $x_1, \ldots, x_n$: $y_n = max\{x_1, \ldots, x_n\}$, so

$$F_{Y_n}(y) = P(Y_n < y) = \prod_{j=1}^{n} P(x_j < y) = F_X(y)^n.$$

The probability density of the maximum y_n is:

$$p_{Y_n}(y) = \frac{dF_X(y)^n}{dy} = nF_X(y)^{n-1} p_X(y).$$

As an example, consider the case in which $x_1, \ldots, x_n$ are uniformly distributed between 0 and X_M: $F_X(x) = x/X_M = 1 - (1 - x/X_M)$ where $0 < x < X_M$. We then have

$$p_{Y_n}(y) = \frac{n}{X_M}\left[1 - \left(1 - \frac{y}{X_M}\right)\right]^{n-1},$$

if we introduce a new variable

$$z = \frac{1}{n}\left(1 - \frac{y}{X_M}\right)$$

in the limit $n \to \infty$ we obtain

$$p_Z(z) = e^{-z}.$$

Let us repeat the previous calculation in the case $F_X(x) = 1 - e^{-\lambda x}, \quad p_X(x) = \lambda e^{-\lambda x}$:

$$F_{Y_n}(y) = \left(1 - e^{-\lambda y}\right)^n,$$

in the limit $n \gg 1$ we have

$$F_{Y_n}(y) \sim e^{-ne^{-\lambda y}}.$$

If we introduce the variable

$$z = \lambda y + \ln(n),$$

we obtain

$$F_Z(z) = e^{-e^{-z}}. \tag{10.20}$$

This distribution is called the Gumbel distribution and appears very often in the statistics of extreme events.

In the two cases previously discussed, we have seen how with a suitable change of variables:

$$x \to z = a_n x + b_n$$

for $n \gg 1$ one obtains a distribution function independent of n, although dependent on $p_X(x)$.

The Gumbel distribution describes the statistics of extreme events under very general assumptions, that is, for all those probability densities $p_X(x)$ such that for large x:

$$\frac{p'_X(x)}{p_X(x)} \sim \frac{p''_X(x)}{p'_X(x)} \sim \frac{p'''_X(x)}{p''_X(x)}. \tag{10.21a}$$

For simplicity, we limit ourselves to discussing the case

$$\frac{p'_X(x)}{p_X(x)} \to const. \tag{10.21b}$$

The strategy consists in introducing a pair of parameters dependent on $p_X(x)$ and then a "natural" variable:

(a) the first parameter is u_n defined by the relation

$$F_X(u_n) = 1 - \frac{1}{n}, \tag{10.22}$$

where n is the number of observations; u_n is the value of x for which the mean return time is n;

(b) the second parameter α_n is given by

$$\alpha_n = n p_X(u_n). \tag{10.23}$$

Eliminating n from the two previous equations we have

$$\alpha_n = \frac{p_X(u_n)}{1 - F_X(u_n)}.$$

In the limit $u_n \to \infty$, we have $p_X(u_n) \to 0$ and $F_X(u_n) \to 1$, using L'Hôpital's rule we obtain

$$\alpha_n \simeq -\frac{p'_X(u_n)}{p_X(u_n)}.$$

The two parameters u_n and α_n allow us to define a new "natural" variable:

$$z = \alpha_n (x - u_n).$$

Note that this procedure is analogous to the one used in the Central Limit Theorem: u_n and α_n play a role analogous to the mean and the inverse of the variance. Let us now show how to derive the Gumbel distribution. Let us consider the identity

$$\ln[n(1 - F_X(x))] = \ln[n(1 - F_X(u_n) + F_X(u_n) - F_X(u_n + (x - u_n)))],$$

using (10.23) we have

$$F_X(u_n + (x - u_n)) - F_X(u_n) \simeq p_X(u_n)(x - u_n) = \frac{p_X(u_n)}{\alpha_n} z = \frac{z}{n},$$

and recalling that $1 - F_X(u_n) = 1/n$ we obtain

$$\ln[n(1 - F_Y(x))] \simeq \ln(1 - z + \ldots) \simeq -z,$$

from which

$$F_X(x) \simeq 1 - \frac{1}{n} e^{-z}.$$

Therefore, for the variable $z = \alpha_n(x - u_n)$ in the limit $n \gg 1$ we have the Gumbel distribution:

$$F_Z(z) = F_X(x)^n \simeq e^{-e^{-z}}.$$

Among the probability densities $p_X(x)$ that satisfy (10.21b) and thus for which extreme events are described (with an appropriate change of variables) by the Gumbel distribution, we mention the gamma distribution

$$p_X(x) = C_a x^a e^{-\lambda y} \quad a > 0 \ \lambda > 0$$

and the logistic distribution

$$p_X(x) = \frac{a}{1 + e^{bx}}.$$

The result we have derived in the particular case of (10.21b) $p'_X(x)/p_X(x) \to const.$ is also valid for a wide class, which includes the Gaussian and the lognormal, for which (10.21a) holds.

10.7 Distribution of Prime Numbers: An Application (at the Limit of What Is Allowed) of Probability Theory

Let us now discuss an example of how probability theory, even in its elementary form, can be useful in branches of mathematics that appear to be very distant.

Let us ask ourselves how the prime numbers are distributed, that is, how many prime numbers are contained between two given numbers n and $n + \Delta n$, or equivalently how many prime numbers are less than a given n. The probabilistic argument we present was devised by Schroeder when he was a young student, and despite its simplicity provides a very accurate answer.

Let us begin by noting that the probability that an integer is divisible by p_j (where p_j is a prime number) is $1/p_j$. In fact, one out of every two numbers is divisible by 2, one out of every three numbers is divisible by 3, one out of every five numbers is divisible by 5, and so on.[6] Assuming that divisibility is an independent property, the probability that x is prime is

$$p(x) = \left(1 - \frac{1}{2}\right)\left(1 - \frac{1}{3}\right)\ldots = \prod_{p_j<x}\left(1 - \frac{1}{p_j}\right). \tag{10.24}$$

Strictly speaking, it would be enough to limit the product to $p_j < \sqrt{x}$, but this is inessential and does not change the result significantly, at least for large x.

Taking the logarithm we have

$$\ln p(x) = \sum_{p_j<x} \ln\left(1 - \frac{1}{p_j}\right) \simeq -\sum_{p_j<x} \frac{1}{p_j}. \tag{10.25}$$

Note now that in the previous equation in the sum each term $1/p_j$ appears with the frequency given by the probability that p_j is prime $p(p_j)$, so we can rewrite (10.25) in the form

$$\ln p(x) = -\sum_{n=2}^{x} \frac{p(n)}{n} \simeq -\int_2^x \frac{p(x')}{x'} dx',$$

differentiating with respect to x we have:

$$\frac{p'(x)}{p(x)} = -\frac{p(x)}{x}. \tag{10.26}$$

Introducing the variable $A(x) = 1/p(x)$, which can be interpreted as the average distance between two prime numbers around x, from (10.26) we have

$$\frac{d}{dx}A(x) = \frac{1}{x}$$

[6] Obviously, we must interpret what we are doing cum grano salis since a number is either prime or it is not.

from which $A(x) = \ln x$ and therefore

$$p(x) = \frac{1}{\ln x}. \tag{10.27}$$

Since the integral of $p(x)$ is divergent, $p(x)$ is not a "true" probability distribution. However, the quantity

$$\Pi(x) = Number\ of\ primes\ less\ than\ x$$

can be expressed as

$$\Pi(x) = \int_2^x p(x')dx'.$$

With the estimate (10.27) we have

$$\Pi(x) = \int_2^x \frac{1}{\ln x'}dx' \simeq \frac{x}{\ln x} \tag{10.28}$$

which constitutes a very accurate approximation.

The procedure we have presented obviously does not claim to be rigorous, but it suggests the correct solution and gives indications for the rigorous proof. It has in fact been proven that

$$\lim_{x\to\infty} \frac{\Pi(x)\ln x}{x} = 1.$$

Let us give some examples of the accuracy of (10.28): for $n = 10^3$, $n = 10^5$ and $n = 10^9$ the exact values of $\Pi(n)$ are 168, 9592 and 50,847,534 to be compared with the approximation (10.28) 177, 9629 and 50,849,234.

Suggested Readings

Apart from the last three sections, the material in this chapter is standard, so it is not necessary to give further indications.

For an introduction to genetics and probability:

W.F. Bodmer, L.L. Cavalli-Sforza, *Genetics of Human Population* (W.H. Freeman, San Francisco, 1978)

The article by Hardy:

G.H. Hardy, Mendelian proportions in a mixed population. Science **28**, 49 (1908)

For a brief discussion on the statistics of extreme events see Chap. 8 of the book

E.W. Montroll, W.W. Badger, *Introduction to Quantitative Aspects of Social Phenomena* (Gorgon and Breach Science Publ., London, 1974)

For number theory and its connections with probability see Chap. 4 of the book

M.R. Schroeder, *Number Theory in Science and Communication* (Springer, New York, 1984)

Solutions

Exercises from Chapter 1

1.1 Make a drawing and (1.12) and (1.13) will be self-evident.

If you want to proceed formally, note that we can write $A \cup B$ in the form $C_1 \cup C_2 \cup C_3$ where $C_1 = A - (A \cap B)$, $C_2 = (A \cap B)$, $C_3 = B - (A \cap B)$. Since these three sets are non-overlapping, we have $P(A \cup B) = P(C_1) + P(C_2) + P(C_3)$, and at this point, since C_1 and $A \cap B$ are disjoint and $A = C_1 \cup (A \cap B)$, then $P(C_1) = P(A) - P(A \cap B)$, similarly $P(C_3) = P(B) - P(A \cap B)$, from which (1.12) follows. In exactly the same way, (1.13) is obtained.

1.2 From (1.12) we have that $P(A \cap B) = P(A) + P(B) - P(A \cup B)$, since $P(A \cup B) \leq 1$ it follows that $P(A \cap B) \geq 3/4 + 1/3 - 1 = 1/12$.

1.3 Using (1.12) we have

$$P((\Omega - A) \cap (\Omega - B)) = P(\Omega - A) + P(\Omega - B) - P((\Omega - A) \cup (\Omega - B))$$

since $P((\Omega - A) \cup (\Omega - B)) \leq 1$, $P(\Omega - A) = 1 - P(A)$ and $P(\Omega - B) = 1 - P(B)$, we have $P((\Omega - A) \cap (\Omega - B)) > 1 - P(A) - P(B) = 1 - 4/10 - 3/10 = 3/10$ therefore there cannot exist two sets A and B that satisfy the requirements.

1.4 Using the notation

$$C_{N,k} = \frac{N!}{k!(N-k)!},$$

the number of possible draws is $C_{90,5}$. To calculate the probability of getting an "ambo", that is, that two given numbers (n_1, n_2) are drawn, we need to determine the number of favorable cases of the type $(n_1, n_2, m_1, m_2, m_3)$ where the numbers

G. Boffetta, A. Vulpiani, *Probability in Physics*, UNITEXT for Physics,
https://doi.org/10.1007/978-3-032-10407-6

m_1, m_2, and m_3 are drawn from the remaining 88 numbers, so we have

$$P(ambo) = \frac{C_{88,3}}{C_{90,5}} = \frac{2}{801},$$

similarly

$$P(single\ number\ drawn) = \frac{C_{89,4}}{C_{90,5}} = \frac{1}{18},$$

$$P(terno) = \frac{C_{87,2}}{C_{90,5}} = \frac{1}{11748},$$

$$P(quaterna) = \frac{C_{86,1}}{C_{90,5}} = \frac{1}{511038}.$$

Exercises from Chapter 2

2.1

(a) Let $P = 1/6$ be the probability of getting a 6 in one roll, there are two possibilities: 6 on the first roll or on the second, the desired probability is $2(1-P)P$.
(b) The probability of getting an even number in one roll is $p = 1/2$, the probability that both numbers are even is p^2.
(c) There are 3 independent possibilities $(1, 3)$, $(2, 2)$ and $(3, 1)$ so the probability that the sum of the numbers is 4 is $3P^2$.
(d) The sum of the numbers is a multiple of 3 if it is 3, 6, 9, or 12; you can get 3 in 2 ways $(1, 2)$ and $(2, 1)$; 6 can occur in 5 ways $(1, 5)$, $(2, 4)$, $(3, 3)$, $(4, 2)$, and $(5, 1)$; 9 can occur in 4 ways $(3, 6)$, $(4, 5)$, $(5, 4)$, and $(6, 3)$; 12 in only one way $(6, 6)$ so the desired probability is $12P^2$.

2.2 Let $P = 1/2$ be the probability of getting heads in one toss

(a) $(1-P)^{N-1}P$
(b) $C_{N,N/2}P^{N/2}(1-P)^{N/2}$
(c) $C_{N,2}P^2(1-P)^{N-2}$
(d) $1 - C_{N,0}(1-P)^N - C_{N,1}P(1-P)^{N-1} = 1 - (1-P)^N - NP(1-P)^{N-1}$.

2.3 Let $P_I = 2/3$ and $P_{II} = 1/3$ be, respectively, the probabilities that a part is produced by industry I or II, with $P(d|I) = 1/5$ and $P(d|II) = 1/50$ the conditional probabilities of getting a defective part from industry I or II.

(a) Using the law of total probability

$$P(a\ defective\ part) = P(d) = P(d|I)P_I + P(d|II)P_{II}.$$

(b) Using Bayes' formula

$$P(I|d) = P(d|I)\frac{P_I}{P(d)}.$$

2.4 Since $P(X = k) = P(Y = k) = 2^{-k}$ we have

$$(a)\quad P(X = Y) = \sum_{k=1}^{\infty} P(X = k)P(Y = k) = \sum_{k=1}^{\infty} 2^{-2k} = \frac{1}{3},$$

(b) We note that $P(X < Y) = P(Y < X)$ and $P(X < Y) + P(Y < X) + P(X = Y) = 1$ from the previous result, so $P(X < Y) = 1/3$

$$(c)\quad P(X\ triple\ of\ Y) = \sum_{k=1}^{\infty} P(X = 3k)P(Y = k) = \sum_{k=1}^{\infty} 16^{-k} = \frac{1}{15}.$$

2.5 The probability that $n - 1$ girls and one boy are born on the n-th attempt is $(1 - P)^{n-1}P$; the average number of boys is 1, in fact for any P we have

$$< n_m >= \sum_{n=1}^{\infty}(1 - P)^{n-1}P = 1.$$

The average number of girls is

$$< n_f >= \sum_{n=1}^{\infty}(n - 1)(1 - P)^{n-1}P = \frac{1 - P}{P},$$

for $P = 1/2$, $< n_f >= 1$.

2.6 The probability that $Y_n < y$ is

$$F_{Y_N}(y) = \prod_{n=1}^{N} P(x_n < y) = F_X(y)^N$$

therefore

$$p_{Y_N}(y) = N F_X(y)^{N-1} p_X(y).$$

Similarly, the probability that $Z_n < z$ is

$$F_{Z_N}(z) = 1 - \prod_{n=1}^{N} P(x_n > z) = 1 - \left(1 - F_X(z)\right)^N$$

from which

$$p_{Z_N}(z) = N\left(1 - F_X(z)\right)^{N-1} p_X(z).$$

2.7 We use polar variables

$$r = \sqrt{x_1^2 + x_2^2}, \ \theta = atang\, z = atang\, \frac{x_2}{x_1},$$

since

$$p_{X_1,X_2}(x_1, x_2) = \frac{1}{2\pi} e^{-(x_1^2+x_2^2)/2}$$

and also $dx_1 dx_2 = r dr d\theta$, the density of variables for (r, θ) is

$$p_{r,\theta}(r, \theta) = \frac{1}{2\pi} r e^{-r^2/2}$$

therefore

$$p_r(r) = r e^{-r^2/2}, \quad p_\theta(\theta) = \frac{1}{2\pi}.$$

Using the change of variables from θ to $z = tang\, \theta$, note that in $\theta \in [-\pi/2, \pi/2]$ so that the transformation is invertible, since $dz/d\theta = 1 + z^2$, we have

$$p_z(z) = \frac{1}{\pi(1 + z^2)}.$$

2.8 Since z and q are linear combinations of Gaussian variables, they are still Gaussian, it is immediate to verify that they are uncorrelated, therefore, since Gaussian variables are also independent.

2.9 Use the result from Exercise 2.7.

2.10 Let X and Y be the arrival times (in minutes), there is a meeting if $|x - y| < 5$ so, since the probability is uniform

$$P(meeting) = \frac{1}{60^2} \int \int_{|x-y|<5} dxdy$$

where x and y vary between 0 and 60 so $P(meeting) = 1 - (11/12)^2 = 23/144$.

2.11 Note that $G(s)$ is the real part of an analytic function. Remember Abel's theorem on power series.

2.12 The variable $Z = X + Y$ is Poissonian with parameter $\lambda_z = \lambda_x + \lambda_y$. Since Y and X are independent $P(X = k, X + Y = n) = P(X = k)P(Y = n - k)$. Recalling that

$$P(X = k) = e^{-\lambda_x} \frac{\lambda_x^k}{k!}, \; P(Y = k) = e^{-\lambda_y} \frac{\lambda_y^k}{k!}, \; P(Z = n) = e^{-\lambda_z} \frac{\lambda_z^n}{n!},$$

we have

$$P(X = k|X + Y = n) = \frac{P(X = k)P(Y = n - k)}{P(Z = n)} =$$

$$\frac{n!}{(n-k)!k!} \left(\frac{\lambda_x}{\lambda_x + \lambda_y}\right)^k \left(1 - \frac{\lambda_x}{\lambda_x + \lambda_y}\right)^{n-k}$$

that is, X, conditioned on $X + Y = n$, has a binomial distribution.

2.13 The probability of getting a 6 in a roll is $P = 1/6$, the probability of getting a 6 for the first time on the n-th roll is $P_n = (1 - P)^{n-1} P$ and the average number is

$$< n > = \sum_{n=1}^{\infty} n(1 - P)^{n-1} P = \frac{1}{P} = 6.$$

2.14 Consider the result of the previous exercise. On the first attempt you surely get a number, afterwards the probability of getting a different number is 4/5, so the average number of boxes to buy is 5/4, for the third number the average number is 5/3 and so on. The average number of boxes to buy to get a prize is therefore

$$1 + \frac{5}{4} + \frac{5}{3} + \frac{5}{2} + 5 = 5\left(\frac{1}{5} + \frac{1}{4} + \frac{1}{3} + \frac{1}{2} + 1\right).$$

In the case with N stickers we have

$$N \sum_{k=1}^{N} \frac{1}{k},$$

in the limit $N \gg 1$ the approximation $N(\ln N + \gamma)$ holds, where $\gamma = 0.57721..$ is the Euler-Mascheroni constant.

2.15 Using the properties of the Dirac delta we have:

$$p_Z(z) = \int\int p_X(x)p_Y(y)\delta(z - xy)dydx = \int\int p_X\Big(\frac{z}{y}\Big)p_Y(y)\frac{dy}{|y|}$$

$$p_Q(q) = \int\int p_X(x)p_Y(y)\delta\Big(q - \frac{x}{y}\Big)dydx = \int\int p_X(qy)p_Y(y)|y|dy.$$

2.16 An explicit calculation shows that $P(X = i) = P(Y = i) = 1/3$ for each i and moreover

$$\langle XY \rangle = 0, \ \langle X \rangle = 0, \ \langle Y \rangle = 0.$$

Note that

$$P(X = 0, Y = 0) = \frac{1}{3} \neq P(X = 0)P(Y = 0) = \frac{1}{9}.$$

Exercises of Chapter 3

3.1 Two estimates can be given using Chebyshev's inequality (CI), or the CLT. Let x_n be the random variable that takes the value 1 (when the needle crosses a line) or 0 with probability respectively p and $1 - p$, of course

$$< x >= p, \ \ \sigma_x^2 = p(1 - p)$$

we denote by $F_N = x_1 + x_2 + \ldots + x_N$ the number of times the needle crosses. From CI we have

$$P\Big(\Big|1 - \frac{F_N}{Np}\Big| > \epsilon\Big) \leq \frac{(1 - p)}{\epsilon^2 Np}.$$

Therefore, if we want a probability less than δ that the percentage error is no more than ϵ, the minimum N is obtained by imposing

$$\frac{(1 - p)}{\epsilon^2 Np} = \delta$$

thus

$$N = \frac{(1-p)}{\epsilon^2 \delta p}.$$

In our case $N = (1-p)10^9/p$, a decidedly large number.

The previous estimate is pessimistic; using the CLT for $N \gg 1$ we have

$$P\Big(\Big|1 - \frac{F_N}{Np}\Big| > \epsilon\Big) = P\Big(\Big|\frac{F_N - Np}{\sqrt{p(1-p)N}}\Big| > \frac{\epsilon p \sqrt{N}}{\sqrt{p(1-p)}}\Big) \simeq 1 - 2\Phi\Big(\frac{\epsilon p \sqrt{N}}{\sqrt{p(1-p)}}\Big)$$

where

$$\Phi(x) = \frac{1}{\sqrt{2\pi}} \int_0^x e^{-z^2/2} dz.$$

The desired value is therefore

$$N = \frac{A^2(1-p)}{\epsilon^2 p}$$

where A is determined by the equation

$$1 - 2\Phi(A) = \delta$$

in our case $\Phi(A) = 0.4995$, from numerical tables we have $A \simeq 3.3$ so

$$N \simeq \frac{(1-p)}{p} 10^7$$

decidedly smaller than the estimate obtained with CI.

Note that the dependence of A (and thus of N) on δ is very weak, as can be seen from the asymptotic behavior for large x of $\Phi(x)$:

$$\Phi(x) \simeq \frac{1}{2} - \frac{1}{\sqrt{2\pi} x} e^{-x^2/2} \Big(1 - \frac{1}{x^2} + \frac{3}{x^4} + \ldots\Big).$$

On the contrary, using CI, N is proportional to δ^{-1}.

3.2 Use the convolution formulas for the sum of independent variables.

3.3 Let us calculate the characteristic function of x_N

$$\phi_{x_N}(t) = \prod_{j=1}^{N} \Big[\frac{1 + e^{it2^{-j}}}{2}\Big],$$

noting that

$$1+e^{it2^{-j}} = \frac{1-\left(e^{it2^{-j}}\right)^2}{1-e^{it2^{-j}}} = \frac{1-e^{it2^{-(j-1)}}}{1-e^{it2^{-j}}},$$

we can write

$$\phi_{x_N}(t) = \frac{1}{2^N}\prod_{j=1}^{N}\left[\frac{1-e^{it2^{-(j-1)}}}{1-e^{it2^{-j}}}\right] = \frac{1}{2^N}\left[\frac{1-e^{it}}{1-e^{it2^{-N}}}\right]$$

in the limit $N \gg 1$

$$\phi_{x_N}(t) \simeq \frac{1-e^{it}}{-it}$$

which is the characteristic function of the uniform probability density in [0, 1].

In the general case, with a completely analogous calculation, we have:

$$\phi_{x_N}(t) = \frac{1}{M^N}\prod_{j=1}^{N}\left[\sum_{k=0}^{M-1} e^{itkM^{-j}}\right] = \frac{1}{M^N}\prod_{j=1}^{N}\left[\frac{1-e^{itM^{-(j-1)}}}{1-e^{itM^{-j}}}\right]$$
$$= \frac{1}{M^N}\left[\frac{1-e^{it}}{1-e^{itM^{-N}}}\right]$$

in the limit $N \gg 1$

$$\phi_{x_N}(t) \simeq \frac{1-e^{it}}{-it}.$$

3.4 The (easy) calculation of the characteristic function of the Gaussian variable x_k with mean m_k and variance σ_k^2 gives

$$\phi_{x_k} = e^{im_k t - \sigma_k^2 t^2/2}$$

using independence, we have

$$\phi_y = e^{iM_N t - S_N^2 t^2/2}$$

where

$$M_N = \sum_{k=1}^{N} m_k \ , \quad S_N = \sum_{k=1}^{N} \sigma_k^2$$

from which the result follows.

3.5 Make a change of variables:

$$x_j = \sum_n C_{jn} z_n$$

so that the $\{z_n\}$ have zero mean and are independent: $< z_n z_j >= \delta_{jn}$. This is always possible since the matrix $A_{ij} =< x_i x_j >$ is symmetric, we have

$$y = \sum_{j,n} C_{jn} z_n$$

and one can use the result from the previous exercise, so y is Gaussian with

$$< y >= 0 \quad < y^2 >=< (\sum_n x_n)^2 >= \sum_n < x_n^2 > +2 \sum_{j<k} < x_j x_k >$$

from which the result follows.

3.6 The calculation of the volume of phase space with energy less than E reduces to that of the volume of a sphere of radius $\sqrt{2mE}$ in $d = 2N$ dimensions. Recalling that the volume of a d-dimensional sphere of radius R is $const.\ R^d$ (the value of the constant is inessential) we have

$$\Sigma(E) = \int_{H<E} d^{2N}\mathbf{q} d^{2N}\mathbf{p} = const.\ L^{2N} (\sqrt{2mE})^{2N} = CE^N$$

the value of C is not important:

$$\omega(E) = \frac{\partial \Sigma(E)}{\partial E} = const.\ E^{N-1}.$$

Since

$$p(E) = const.\, \omega(E) e^{-\beta E}$$

we have

$$p(E) = \frac{E^{N-1}e^{-E/k_BT}}{\int_0^\infty E^{N-1}e^{-E/k_BT}dE} = \frac{E^{N-1}e^{-E/k_BT}}{(k_BT)^N\Gamma(N)},$$

which reaches its maximum at $E^* = (N-1)k_BT$.

The calculation of the moments $< E^n >$ is immediate (just recall the definition of the Euler gamma function):

$$< E^n >= \frac{\int_0^\infty E^{N+n-1}e^{-E/k_BT}}{(k_BT)^N\Gamma(N)} = (k_BT)^n\frac{\Gamma(N+n)}{\Gamma(N)}, \quad < E >= Nk_BT,$$

from which the result follows.

3.7 With a suitable change of variables:

$$x_j = \sum_n C_{jn}z_n$$

the Hamiltonian is diagonalized, note that this is always possible since the matrix A_{ij} is symmetric, so we have

$$H = \sum_{n=1}^{N}\left[\frac{p_n^2}{2m} + a_n x_n^2\right]$$

where $\{a_n\}$ are the positive eigenvalues of the matrix $\{A_{n,j}\}$, at this point the problem is completely analogous to the previous one.

3.8 Since the system is in thermal equilibrium at temperature T, the joint probability density of the positions of the light particles and the heavy one is

$$p(x_1, x_2, \dots, x_N, X) = const. e^{-\beta V(x_1, x_2, \dots, x_N, X)}$$

where

$$V(x_1, x_2, \dots, x_N, X) = \begin{cases} FX \ \ if \ \ for \ \ every \ \ n \ \ x_n \in [0, X] \\ \infty \ \ otherwise \end{cases}$$

we therefore have

$$p(x_1, x_2, \dots, x_N, X) = const. \prod_{n=1}^{N}\theta(X - x_n)e^{-\beta FX}$$

where $\theta(\ )$ is the step function. To obtain the (marginal) probability density of X, one must integrate over $x_1, x_2, \ldots, x_N$:

$$p(X) = const.\, X^N e^{-\beta F X} = \frac{(\beta F)^{N+1}}{N!} X^N e^{-\beta F X},$$

where the normalization constant is determined using gamma functions.

For case (b), in the calculation of the new probability density $\tilde{p}(X)$ one must take into account the constraint that one of the $\{x_n\}$ is zero, so

$$\tilde{p}(X) = const. \int \prod_{n=1}^{N} \theta(X - x_n) \sum_{n=1}^{N} \delta(x_n)\, e^{-\beta F X}\, dx_1, dx_2, \ldots dx_N =$$

$$\frac{(\beta F)^N}{(N-1)!} X^{N-1} e^{-\beta F X}.$$

3.9 A simple calculation in complex analysis shows that

$$\phi_{x_j}(t) = \int_{-\infty}^{\infty} \frac{e^{itx}}{\pi(1+x^2)} dx = e^{-|t|},$$

since

$$\phi_y(t) = \left[\phi_x(t/N)\right]^N = e^{-|t|},$$

we have that the probability density of the variable Y is

$$p_Y(y) = \frac{1}{\pi(1+y^2)}.$$

Since the variance of the $\{x_j\}$ is infinite, the law of large numbers cannot be used.

3.10 Once two series of i.i.d. random numbers uniformly distributed in (0, 1) are generated

$$x_1, x_2, \ldots, x_N, \quad y_1, y_2, \ldots, y_N$$

we introduce the variables $\{i_n\}$:

$$i_n = \begin{cases} 1 & if\ y_n < f(x_n) \\ 0 & if\ y_n > f(x_n) \end{cases}$$

The quantity

$$I_N = \frac{1}{N} \sum_{n=1}^{N} i_n$$

is the fraction of points that fall between 0 and the curve $f(x)$ and, by the law of large numbers, in the limit $N \gg 1$ tends to

$$\int_0^1 f(x)dx.$$

For an estimate of the convergence of I_N and the error, proceed as in exercise 3.1.

3.11 Let us consider $n_1, n_2, \ldots, n_N$ i.i.d. Poisson variables with

$$P(n_j = k) = \frac{\lambda^k}{k!} e^{-\lambda}$$

the variable $S_N = n_1 + n_2 + \ldots + n_N$ is still Poisson with

$$P(S_N = k) = \frac{(\lambda N)^k}{k!} e^{-\lambda N}.$$

Therefore, if $\lambda = 1$

$$P(S_N \leq N) = e^{-N}\Big(1 + N + \frac{N^2}{2!} + \frac{N^3}{3!} + \ldots + \frac{N^N}{N!}\Big).$$

If $N \gg 1$ then, since $< n_j >= \sigma_{n_j} = 1$, by the CLT the previous probability can be calculated as follows

$$\lim_{N\to\infty} P(\frac{S_N - N}{\sqrt{N}} \leq 0) = \frac{1}{\sqrt{2\pi}} \int_{-\infty}^{0} e^{-x^2/2} dx = \frac{1}{2},$$

from which the first result.

Similarly, we have

$$P(x_1 \leq \frac{S_N - N}{\sqrt{N}} \leq x_2) = e^{-N} \sum_{k=[N+x_1\sqrt{N}]}^{[N+x_2\sqrt{N}]} \frac{N^k}{k!}.$$

In the limit $N \to \infty$ we can use the CLT and thus

$$\lim_{N\to\infty} e^{-N} \sum_{k=[N+x_1\sqrt{N}]}^{[N+x_2\sqrt{N}]} \frac{N^k}{k!} = \frac{1}{\sqrt{2\pi}} \int_{x_1}^{x_2} e^{-x^2/2} dx.$$

Exercises of Chapter 4

4.1 Proceeding as in Sect. 4.4 we have that

$$< v_n >=< v_0 > a^n, \quad < v_{n+k} v_n >=< v^2 > a^k, \quad < v_{n+1}^2 >= a^2 < v_n^2 > +\sigma_z^2$$

therefore in the limit $n \to \infty$

$$< v >= 0, \quad < v^2 >= \frac{\sigma_z^2}{1-a^2}.$$

(a) Denoting by $p_V(v, n)$ the probability density of v_n we have

$$\begin{aligned} p_V(v, n+1) &= \int\int p_V(v', n) p_Z(z) \delta(v - av' - z) dv' dz \\ &= \int p_V\Big(\frac{v-z}{a}, n\Big) p_Z(z) \frac{dz}{|a|}, \end{aligned}$$

assuming that for $n \to \infty$, $p_V(v, n) \to p_V(v)$ we obtain

$$p_V(v) = \int\int p_V(v') p_Z(z) \delta(v - av' - z) dv' dz = \int p_V\Big(\frac{v-z}{a}\Big) p_Z(z) \frac{dz}{|a|},$$

which has a Gaussian solution only if $p_Z(z)$ is Gaussian. This is evident if we write the previous equation in terms of characteristic functions:

$$\phi_V(t) = \phi_V(at) \phi_Z(t), \quad \ln \phi_V(t) = \ln \phi_V(at) + \ln \phi_Z(t),$$

if z is not a Gaussian variable then $\ln \phi_Z(t) \neq -const.t^2$, also $\phi_V(t) \neq -const.t^2$ and therefore v is not Gaussian.

(b) We write

$$x_n - x_0 = \sum_{j=0}^{n-1} v_j,$$

the $\{v_j\}$ are weakly correlated so the CLT holds, a calculation entirely analogous to that done in Chap. 4 shows

$$D =< v^2 > \Big(\frac{1}{2} + \sum_{j=1}^{\infty} a^j\Big) =< v^2 > \Big(\frac{1}{2} + \frac{a}{1-a}\Big).$$

4.2 Since $x_{n+1} = ax_n + z_n$ where x_n and z_n are independent, denoting respectively by $\phi_n(t)$ and $\phi_z(t)$ the characteristic functions of x_n and z, we have

$$\phi_{n+1}(t) = \phi_n(at)\phi_z(t), \;\; \ln \phi_{n+1}(t) = \ln \phi_n(at) + \ln \phi_z(t).$$

Indicating with c_k and $C_k^{(n)}$ the cumulants of order k of z and x_n, respectively, we have:

$$\sum_{k=1} \frac{C_k^{(n+1)}}{k!}(it)^k = \sum_{k=1} \frac{C_k^{(n)}}{k!}(iat)^k + \sum_{k=1} \frac{c_k}{k!}(it)^k,$$

from which the recurrence rule

$$C_k^{(n+1)} = a^k C_k^{(n)} + c_k$$

which is easily solved:

$$C_k^{(n)} = C_k^{(\infty)} + a^{kn}\left(C_k^{(0)} - C_k^{(\infty)}\right), \;\; C_k^{(\infty)} = \frac{c_k}{1-a^k},$$

we therefore have:

$$\lim_{n\to\infty} \ln \phi_{x_n}(t) = \sum_{k=1} \frac{c_k}{k!}\left[\frac{1}{1-a^k}\right](it)^k.$$

In the case where z is a Gaussian with zero mean, we obtain the result seen in Sect. 4.4 for the discrete-time model of Brownian Motion.

The relaxation time of the k-th cumulant is

$$n_k = -\frac{1}{\ln |a^k|} = -\frac{1}{k \ln |a|},$$

the relaxation time of $p_{x_n}(x)$ is determined by the longest time, thus n_1 if $c_1 = \langle z \rangle \neq 0$ or $n_2 = n_1/2$ if $\langle z \rangle = 0$.

Exercises of Chapter 5

5.1 Let us write the equation for the evolution of the probability of being in state 1 at time $t + 1$:

$$P_1(t+1) = P_1(t)P_{1\to 1} + P_2(t)P_{2\to 1}.$$

Noting that $P_2(t) = 1 - P_1(t)$ we have

$$P_1(t+1) = (1-a-b)P_1(t) + b$$

which is easily solved:

$$P_1(t) = \frac{b}{a+b} + (1-a-b)^t \Big[P_1(0) - \frac{b}{a+b} \Big]$$

thus for $t \to \infty$

$$(P_1(t), P_1(t)) \to (\pi_1, \pi_2)$$

where

$$(\pi_1, \pi_2) = \Big[\frac{b}{a+b}, \frac{a}{a+b} \Big].$$

The characteristic relaxation time is

$$\tau_c = \frac{-1}{\ln |1-a-b|}$$

which diverges when $|1-a-b| \to 1$, for example $a \to 0$ and $b \to 0$; or $a \to 1$ and $b \to 1$.

The detailed balance $\pi_1 P_{1\to 2} = \pi_2 P_{2\to 1}$ holds for any a and b such that $0 < a < 1, 0 < b < 1$.

5.2 The equations for the invariant probabilities are

$$\pi_k = \sum_{j=1}^{N} P_{j\to k}\pi_j,$$

it is immediate to verify that if

$$\sum_{j=1}^{N} P_{j\to k} = 1 \quad for \ every \ k$$

a solution is $\pi_j = 1/N$, if the chain is ergodic this is the only solution.

5.3 Let us denote by 1 and 2 the states of good weather and bad weather, respectively, we have $P_{1\to 1} = 2/3$ and $P_{2\to 2} = 3/4$, the invariant probabilities are found from exercise *E5.1* with $a = 1/3$ and $b = 1/4$:

$$\pi_1 = \frac{3}{7}, \quad \pi_2 = \frac{4}{7}$$

(a) The average number of good weather days in a year is $365\pi_1 \simeq 156$.
(b) To have 5 consecutive good weather days after a bad weather day, there must be a $2 \to 1$ transition and then 4 transitions $1 \to 1$, so the probability is

$$P_{2\to 1}(P_{1\to 1})^4 = \frac{1}{4}\left(\frac{2}{3}\right)^4.$$

5.4 The transition probabilities are:

$$P_{n\to n} = \frac{1}{2}, \quad P_{n\to k} = \frac{1}{4}, \quad if \quad n \neq k.$$

(a) $\pi_n = 1/3$, this is also evident by symmetry;
(b) if $y_t = I$ then $x_t = 1$ and, for the original variables, there are 3 possible transitions $x_{t+1} = 1$ with probability $1/2$, or 2 or 3 both with probability $1/4$, so, since the $\{x_t\}$ are a Markov chain, the transitions of the variable $\{y_t\}$ starting from state I do not depend on the past and we have

$$P_{I\to I} = \frac{1}{2}, \quad P_{I\to II} = \frac{1}{2}.$$

Now consider the case $y_t = II$, this means that x_t is either 2 or 3, starting from one of these two states there is a probability $1/4$, independent of the past, to transition to $x_{t+1} = 1$ (and thus in $y_{t+1} = I$), instead with probability $3/4$ one remains in II:

$$P_{II\to I} = \frac{1}{4}, \quad P_{II\to II} = \frac{3}{4},$$

therefore $\{y_t\}$ is a Markov chain.

The invariant probabilities are $(\pi_I = 1/3, \pi_{II} = 2/3)$; note that $\pi_I = \pi_1$ and $\pi_{II} = \pi_2 + \pi_3$.

In general, it is NOT true that starting from a Markov chain, by grouping some states, one still obtains a Markov chain, see the following exercise.

5.5 Let us consider the case $y_t = I$, the probability of having $y_{t+1} = I$ or $y_{t+1} = II$ depends on y_{t-1}, in fact if $y_{t-1} = II$ then $y_{t+1} = I$ with probability 1, on the contrary if $y_{t-1} = I$ then $y_{t+1} = I$ with probability different from 1. Therefore, the stochastic process $\{y_t\}$ is not a Markov chain.

5.6 The transition probabilities are

$$P_{n\to n} = 0, \quad P_{n\to k} = \frac{1}{3}, \quad if \ n \neq 4 \ and \ k \neq n,$$

$$P_{4\to 4} = 1, \quad P_{4\to n} = 0 \ if \ n \neq 4.$$

At each step, if one is in a state j different from 4, the probability of not getting trapped is

$$\sum_{n\neq 4} P_{j\to n} = \frac{2}{3},$$

therefore the required probability is $(2/3)^N$.

5.7 Let y_{n+1} be the result obtained on the $n+1$-th roll, we have

$$x_{n+1} = \max(x_n, y_{n+1}).$$

If $x_n = 1$ all transitions are equally probable:

$$P_{1\to j} = \frac{1}{6} \quad for \;\; every \;\; j,$$

if $x_n = 2$, the transition to 1 is impossible, the transition to 2 occurs if y_{n+1} is 1 or 2, the transition to $j \geq 3$ occurs if $y_{n+1} = j$ so

$$P_{2\to j} = \begin{cases} 0 \;\; if \;\; j=1 \\ 1/3 \;\; if \;\; j=2 \\ 1/6 \;\; if \;\; j\geq 3 \end{cases}$$

Similarly,

$$P_{3\to j} = \begin{cases} 0 \;\; if \;\; j=1,2 \\ 1/2 \;\; if \;\; j=3 \\ 1/6 \;\; if \;\; j\geq 4 \end{cases}$$

and so on, therefore

$$P_{i\to j} = \begin{cases} 0 \;\; if \;\; j<i \\ j/6 \;\; if \;\; j=i \\ 1/6 \;\; if \;\; j>i \end{cases}$$

State 6 is absorbing and the invariant probabilities are (0, 0, 0, 0, 0, 1).

5.8 Let us write the equations for the invariant probabilities:

$$\pi_1 = \frac{1}{2}\pi_2, \quad \pi_L = \frac{1}{2}\pi_{L-1}, \quad \pi_2 = \pi_1 + \frac{1}{2}\pi_3, \quad \pi_{L-1} = \pi_L + \frac{1}{2}\pi_{L-2}$$

$$\pi_n = \frac{1}{2}(\pi_{n-1} + \pi_{n+1}), \quad n = 3, 4, \ldots, L-2.$$

The solution is:

$$\pi_1 = \pi_L = \frac{1}{2(L-1)}, \quad \pi_n = \frac{1}{(L-1)}, \quad n = 2, 3, \ldots, L-1.$$

5.9 Let us write the equations for the invariant probabilities:

$$\pi_0 = \frac{\pi_1}{1+r}, \quad \pi_1 = \pi_0 + \frac{\pi_2}{1+r}$$

$$\pi_n = \frac{\pi_{n+1}}{1+r} + \frac{r}{1+r}\pi_{n-1}, \quad n \geq 2.$$

For $r < 1$ the solution is

$$\pi_{n+1} = r\pi_n = r^n \pi_1, \quad n > 1$$

imposing normalization

$$\sum_{n=0}^{\infty} \pi_n = 1$$

one calculates π_1 and π_0:

$$\pi_0 = \frac{1-r}{2}, \quad \pi_n = \frac{1-r^2}{2} r^{n-1}, \quad n \geq 1.$$

5.10 Let us write the equations for the invariant probabilities:

$$\pi_n = \pi_{n+1} + f_n \pi_0,$$

in recursive form:

$$\pi_{n+1} = \pi_n - f_n \pi_0.$$

Looking for a solution of the form

$$\pi_n = c_n \pi_0,$$

we obtain

$$c_{n+1} = c_n - f_n, \quad c_n = 1 - \sum_{k=0}^{n-1} f_k$$

and therefore

$$\pi_n = \left(1 - \sum_{k=0}^{n-1} f_k\right)\pi_0.$$

The probability π_0 is determined by the normalization condition

$$\pi_0 + \sum_{n=1}^{\infty} \pi_n = 1, \quad \pi_0 = \frac{1}{1 + \sum_{n=1}^{\infty} c_n}.$$

If $f_n \sim n^{-\beta}$, recalling that

$$\sum_{k=0}^{\infty} f_k = 1$$

we have

$$c_n = 1 - \sum_{k=0}^{n-1} f_k = \sum_{k=n}^{\infty} f_k \sim n^{-(\beta-1)}$$

and therefore, if $\beta > 2$

$$\sum_n c_n$$

is finite, and normalization is possible.

In the opposite case ($1 < \beta \leq 2$) there are no invariant probabilities: the system "escapes to infinity", see the cited article by Wang.

5.11 The probability $P_{n \to j}$ is given by the coefficient of the power s^j in the series expansion of the generating function $G_n(s)$ corresponding to the system with n individuals. This function is nothing but the n-th power of $G(s)$:

$$G(s) = \sum_{k=0}^{\infty} s^k P_k = \sum_{k=0}^{\infty} \frac{s^k \lambda^k}{k!} e^{-\lambda} = e^{\lambda(s-1)},$$

therefore

$$G(s)^n = e^{n\lambda(s-1)} = \sum_{k=0}^{\infty} \frac{s^k (n\lambda)^k}{k!} e^{-n\lambda},$$

from which

$$P_{n\to j} = \frac{(n\lambda)^j}{j!} e^{-n\lambda}.$$

5.12

(a) Let's calculate the two-step transition probabilities:

$$P^{(2)}_{1\to 1} = 1, \quad P^{(2)}_{2\to 2} = P^{(2)}_{2\to 3} = P^{(2)}_{3\to 2} = P^{(2)}_{3\to 3} = \frac{1}{2}$$

the others are zero; calculating those for three steps we find

$$P^{(3)}_{i\to j} = P_{i\to j},$$

from which we have

$$P^{(2n-1)}_{i\to j} = P_{i\to j}, \quad P^{(2n)}_{i\to j} = P^{(2)}_{i\to j},$$

so the chain is periodic.

(b) Note that if at time zero the system is in state 1, it will never be in this state at odd times and will certainly be in it at even times, thus we have $f_1 = 1/2$, and similarly we obtain $f_2 = f_3 = 1/4$.

Exercises of Chapter 6

6.1 Let's integrate the Langevin equation starting from a given $x(0)$:

$$x(t) = x(0)e^{-t/\tau} + c\int_0^t e^{-(t-t')/\tau}\eta(t')dt'$$

therefore

$$< x(t)x(0) >=< x(0)^2 > e^{-t/\tau} + c\int_0^t e^{-(t-t')/\tau} < x(0)\eta(t') > dt'$$

from which, since $< x(0)\eta(t') >= 0$, the result follows.

6.2 Taking the average of

$$x_{t+1} = ax_t + b\sum_{j=0}^{N-1} 2^{-j} z_{t-j}$$

we have

$$< x_{t+1} >= a < x_t >, \quad < x_t >=< x_0 > a^t .$$

Similarly

$$< x_{t+1}^2 >= a^2 < x_t > +b^2 \sum_{j=0}^{N-1} 4^{-j} = a^2 < x_t > +C, \quad C = 4b^2 \frac{1-4^{-N}}{3}.$$

In the limit $t \to \infty$:

$$< x >= 0, \quad < x^2 >= \sigma_x^2 = \frac{C}{1-a^2}.$$

Introducing the variable

$$Q_{t+1} = b \sum_{j=0}^{N-1} 2^{-j} z_{t-j}$$

we can write, $x_1 = ax_0 + Q_1$, $x_2 = a(ax_0 + Q_1) + Q_2 = a^2 x_0 + aQ_1 + Q_2$, iterating:

$$x_t = a^t x_0 + a^{t-1} Q_1 + a^{t-2} Q_2 + \ldots + a Q_{t-1} + Q_t$$

for large t the first term is negligible, using the result of Exercise 3.5 the result follows.

Exercises of Chapter 7

7.1 Let us consider a sufficiently regular function

$$A(x) = A_0 + \sum_{n \neq 0}^{\infty} A_n e^{2\pi i n x}$$

where A_n decays rapidly to zero for $|n| \to \infty$ (for example $|A_n| < const. C^{|n|}$ with $C < 1$), obviously

$$< A >= A_0.$$

Since $x_t = x_0 + \omega t \ \ mod\ 1$ we can write the time average up to time T as

$$\frac{1}{T}\sum_{t=0}^{T-1} A(x_t) = A_0 + \frac{1}{T}\sum_{t=0}^{T-1}\sum_{n\neq 0}^{\infty} A_n e^{2\pi i n(x_0+\omega t)}$$

using the formula for the sum of geometric series we have

$$\frac{1}{T}\sum_{t=0}^{T-1} A(x_t) = A_0 + \frac{1}{T}\sum_{n\neq 0}^{\infty} A_n e^{2\pi i n x_0}\Big[\frac{1-e^{2\pi i n\omega T}}{1-e^{2\pi i n\omega}}\Big].$$

If ω is irrational, $n\omega$ cannot be an integer, so the term in square brackets is bounded and in the limit $T \to \infty$ the time average tends to $A_0 =< A >$ independently of x_0.

If ω is rational, it can be written in the form $\omega = p/q$ where p and q are integers, so the motion is periodic, in fact $x_{t=q} = x_0 + p \ mod\ 1 = x_0$, and the system is not ergodic.

7.2 Consider the function

$$A(x) = \begin{cases} A_1 & if\ x \in I_1 \\ A_2 & if\ x \in I_2 \end{cases},$$

where $I_1 = [0, 1/4] \cup [3/4, 1]$ and $I_2 = [1/4, 3/4]$. It is easy to see that I_1 is an invariant interval under the dynamics, in fact if $x_0 \in I_1$ then for every t we have $x_t \in I_1$, so the time average of $A(x)$ will be A_1; similarly, if $x_0 \in I_2$ we obtain A_2; while the average of $A(x)$ with respect to the uniform probability density is $(A_1 + A_2)/2$, so the system is not ergodic.

7.3 Let us write the equation for the invariant density. There are 2 preimages of x: $x^{(1)} = xp$ and $x^{(2)} = 1-x(1-p)$, since $f'(x^{(1)}) = 1/p$ and $f'(x^{(2)}) = 1/(p-1)$, we have

$$\rho_{inv}(x) = p\rho_{inv}(x^{(1)}) + (1-p)\rho_{inv}(x^{(2)}),$$

which admits the solution

$$\rho_{inv}(x) = 1.$$

7.4 If $x \in (0, 1/2)$ there are 2 preimages $x^{(1)} = x/2$ and $x^{(2)} = 1/2 + x$, if $x \in (1/2, 1)$ only $x^{(1)} = x/2$, $f'(x^{(1)}) = 2$ and $f'(x^{(2)}) = 1$.

We therefore have

$$\rho_{inv}(x) = \begin{cases} \rho_{inv}(x/2)/2 + \rho_{inv}(1/2+x) & if\ 0 \le x < 1/2 \\ \rho_{inv}(x/2)/2 & if\ \ 1/2 \le x < 1 \end{cases}$$

It is immediate to see that a solution is

$$\rho_{inv}(x) = \begin{cases} 4/3 & if\ 0 \le x_t < 1/2 \\ 2/3 & if\ 1/2 \le x_t < 1 \end{cases}$$

If the problem is studied with a two-state Markov chain, where state 1 corresponds to $x \in [0, 1/2]$ and 2 if $x \in (1/2, 1)$, we have

$$P_{1\to1} = \frac{1}{2},\ \ P_{1\to2} = \frac{1}{2},\ \ P_{2\to1} = 1,\ \ P_{2\to2} = 0,$$

from which

$$\pi_1 = \frac{2}{3},\ \ \pi_1 = \frac{1}{3},$$

note that

$$\pi_1 = \int_0^{1/2} \rho_{inv}(x)dx,\ \ \pi_2 = \int_{1/2}^{1} \rho_{inv}(x)dx.$$

Exercises of Chapter 9

9.1 For clarity, we use the notation $p_X(x)$ and $q_X(x)$ to indicate the probability densities, with a change of variables $x \to y = f(x)$ where the transformation is invertible, we obtain $p_Y(y)$:

$$p_Y(y) = \frac{p_X(x^*)}{|f'(x^*)|} \quad where\ x^* = f^{(-1)}(y)$$

and analogously $q_Y(y)$, note that

$$\frac{p_X(x)}{q_X(x)} = \frac{p_Y(y)}{q_Y(y)},$$

and moreover

$$\tilde{S}[p|q] = -\int p_X(x) \ln\left[\frac{p_X(x)}{q_X(x)}\right]dx = -E\Big(\ln\left[\frac{p_X(x)}{q_X(x)}\right]\Big)$$

from which the result follows.

9.2 Once a positive velocity v is drawn, the crossing time from $x = 0$ to $x = L$ is $\Delta t(v) = L/v$, and this is also the return time.

Let us consider the case with $v > 0$: in a very long time interval T there are $N \gg 1$ draws, by the law of large numbers we have that the fraction of times in which $\Delta t \in [\Delta t(v),\ \Delta t(v - dv)]$ is

$$F(v, dv) = const.\, g(v)\, dv,$$

to calculate the fraction of time in which the velocity is in the interval $[v - dv,\ v]$ one must multiply $F(v, dv)$ by $\Delta t(v) = L/v$, thus we have

$$p_V(v) = const.\, \frac{g(v)}{v}.$$

By symmetry, the same result holds for the case $v < 0$, therefore:

$$p_V(v) = \frac{1}{\sqrt{2\pi}} e^{-v^2/2}.$$

Noting that once a v is drawn, the residence time in $[x - \Delta x,\ x]$ is independent of x and v, we have

$$p_X(x) = \frac{1}{L}.$$

9.3 From Bayes' formula we have

$$P(x_0 = j|x_n = i) = P(x_n = i|x_0 = j)\frac{P(x_0 = j)}{P(x_n = i)},$$

using the notation

$$P(x_0 = j) = \pi_j, \quad P(x_n = i|x_0 = j) = P^{(n)}_{j\to i},$$

we obtain

$$H(x_0|x_n) = -\sum_{i,j} P(x_n = i)P(x_0 = j|x_n = i) \ln P(x_0 = j|x_n = i) =$$

$$-\sum_{i,j} \pi_j P^{(n)}_{j\to i}\Big(\ln P^{(n)}_{j\to i} + \ln \pi_j - \ln \pi_i \Big).$$

The first term

$$-\sum_{i,j} \pi_j P^{(n)}_{j\to i} \ln P^{(n)}_{j\to i}$$

is nothing but $H(x_n|x_0)$, the other terms cancel each other. Indeed, since $\sum_i P^{(n)}_{j\to i} = 1$, we have

$$-\sum_{i,j} \pi_j P^{(n)}_{j\to i} \ln \pi_j = -\sum_j \pi_j \ln \pi_j$$

similarly, noting that $\sum_j \pi_j P^{(n)}_{j\to i} = \pi_i$, we have

$$\sum_{i,j} \pi_j P^{(n)}_{j\to i} \ln \pi_i = \sum_i \pi_i \ln \pi_i,$$

from which the result follows. In the limit $n \gg 1$ we have $P^{(n)}_{j\to i} \simeq \pi_i$ and thus $H(x_0|x_n) = H(x_n|x_0) \simeq -\sum_i \pi_i \ln \pi_i$.

9.4 Using the notation from the previous exercise we can write $H(x_n|x_0 = i)$ in the form

$$H(x_n|x_0 = i) = -\sum_j P^{(n)}_{i\to j} \ln P^{(n)}_{i\to j}.$$

From Bayes' formula

$$P(x_0 = j|x_n = i) = P^{(n)}_{j\to i} \frac{\pi_j}{\pi_i},$$

and from detailed balance

$$P^{(n)}_{i\to j}\pi_i = P^{(n)}_{j\to i}\pi_j$$

we obtain

$$P(x_0 = j|x_n = i) = P^{(n)}_{i\to j}.$$

Therefore,

$$H(x_0|x_n = i) = -\sum_j P(x_0 = j|x_n = i) \ln P(x_0 = j|x_n = i) =$$

$$-\sum_j P^{(n)}_{i\to j} \ln P^{(n)}_{i\to j} = H(x_n|x_0 = i).$$

9.5 Note that there is never a $2 \to 2$ transition, so the words of length N ending with 2 are necessarily generated from those of length $N-1$ ending with 1, whereas the words of length N ending with 1 are generated both from those of length $N-1$

ending with 1, and from those ending with 2. Let $\mathcal{N}_1(N)$ and $\mathcal{N}_2(N)$ be the number of words of length N ending with 1 and 2, respectively, then

$$\mathcal{N}_1(N) = \mathcal{N}_1(N-1) + \mathcal{N}_2(N-1), \quad \mathcal{N}_2(N) = \mathcal{N}_1(N-1).$$

The total number of words of length N is $\mathcal{N}(N) = \mathcal{N}_1(N) + \mathcal{N}_2(N)$, from the previous equation we have

$$\mathcal{N}(N) = \mathcal{N}(N-1) + \mathcal{N}(N-2),$$

which is nothing but the Fibonacci sequence. Assuming that $\mathcal{N}(N) \sim G^N$ we obtain

$$G = 1 + \frac{1}{G}$$

from which $G = (1+\sqrt{5})/2$. Let us compare the number of allowed words $\sim G^N$ with that of typical words (Shannon-McMillan theorem) $\sim e^{h_S N}$ where

$$h_S = -\frac{1}{2-p}\Big[p \ln p + (1-p)\ln(1-p)\Big] \leq \ln G,$$

equality holds for $p = G - 1$, thus (apart from the case $p = G - 1$) we have

$$G^N \gg e^{h_S N}.$$

The manufacturer's authorised representative in the EU is Springer Nature Customer Service Centre GmbH, Europaplatz 3, 69115 Heidelberg, Germany. If you have any concerns regarding our products, please contact ProductSafety@springernature.com

Printed and bound by CPI Group (UK) Ltd, Croydon, CR0 4YY

07/07/2026

02160923-0001